The Problem Solver™

Student Workbook
Second Edition

Judy Goodnow
Shirley Hoogeboom

Wright Group

www.WrightGroup.com

Copyright © 2008 by Wright Group/McGraw-Hill.

All rights reserved. Except as permitted under the United States
Copyright Act, no part of this publication may be reproduced or
distributed in any form or by any means, or stored in a database
or retrieval system, without the prior written permission from the
publisher, unless otherwise indicated.

Printed in the United States of America.

Send all inquiries to:
Wright Group/McGraw-Hill
P.O. Box 812960
Chicago, IL 60681

ISBN 978-0-07-704102-1
MHID 0-07-704102-X

4 5 6 7 8 9 QVS/QVS 16 15 14 13

NAME _______________________

1 FIND OUT

What question do you have to answer?

Find out what the problem tells you.

2 CHOOSE A STRATEGY

I can ________________________
to solve the problem.

  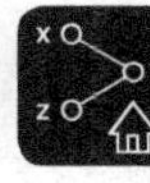  

3 SOLVE IT

Show your work.

4 LOOK BACK

- Read the problem again.
- Check your work.
- Did you answer the question?
- Does your answer make sense?
- How do you know?

Use or Make a Table

1

Tad the Talking Robot can keep anyone company! When the robots are made, every 8th robot to move along the conveyor belt gets a blue control panel, every 3rd robot gets blinking green eyes, and every 4th robot gets a square head. If 150 robots come off the conveyor belt, how many will have all three: a blue control panel, blinking green eyes, and a square head?

① FIND OUT

A. What is the question you have to answer?

B. What is being added to the robots?

C. Which robots get blinking green eyes? A blue control panel? A square head?

② CHOOSE A STRATEGY

I can _________________________________ to solve the problem.

         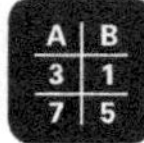

Robot	1	2	3	4	5	6	
Blue Panel							
Green Eyes			X			X	
Square Head				X			

A. What will you keep track of in the first row of the table? In the second row? In the third row? In the fourth row?

B. Which is the first robot to get a blue control panel? The next?

C. Which is the first robot to get blinking green eyes? The next?

D. Which is the first robot to get a square head? The next?

E. Keep filling in the table. Which robot is the first to have all three? What is the next robot to have all three? Do you see a pattern?

F. Use the pattern. How many robots out of 150 will have a blue control panel, blinking green eyes, and a square head?

Read the problem again. Look at the information given and the main question. Review your work. Is your answer reasonable?

Use or Make a Table

2

Lourdes is using a train schedule to help her relatives plan a family reunion. Her relatives will arrive from San Francisco, Los Angeles, Salt Lake City, Denver, and Boston. A train arrives from San Francisco every 4 days, Los Angeles every 10 days, Salt Lake City every 3 days, Denver every 8 days, and Boston every 15 days. Lourdes' relatives are all trying to arrive on the same day. The trains all arrived together today. How many times in the next 20 weeks could the family members all arrive on the same day?

1 FIND OUT

A. What is the question you have to answer?

B. How often does the train come from San Francisco? Los Angeles? Salt Lake City? Denver? Boston?

2 CHOOSE A STRATEGY

I can _______________________________ to solve the problem.

       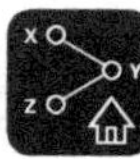  

Boston—15 days	15	30	45	
Los Angeles—10 days	N	Y	N	
Denver—8 days		N		
San Francisco—4 days				
Salt Lake City—3 days				

A. Why does the table begin with the days the train will come in from Boston? What labels are shown for the other rows?

B. How many days are in 20 weeks? Complete the first row. How many times will the train come in from Boston?

C. How often does the train arrive from Los Angeles? Will it arrive on the 15th day? Do you need to check the rest of the trains for that day? What is the first day on which at least two trains will arrive? Complete the Los Angeles row.

D. Will the train from Denver come in on the 30th day? The 45th? Keep looking for a day when all the trains come in.

E. How many times in the next 20 weeks could the family members all arrive on the same day?

4 LOOK BACK

Read the problem again. Look at the information given and the main question. Review your work. Is your answer reasonable?

Make an Organized List

3

Seneca and Artie are lab partners in science class. Today they have to weigh liquid. They have a tray of 80 weights that they can use. There are four different kinds of weights: 50 grams, 25 grams, 15 grams, and 5 grams. The first liquid weighs 85 grams. How many different combinations of weights will balance the scale for the first liquid?

① FIND OUT

A. What is the question you have to answer?

B. What are Seneca and Artie weighing? How many different kinds of gram weights do they have? What are the different kinds?

C. How much does the first liquid weigh?

② CHOOSE A STRATEGY

I can _________________________________ to solve the problem.

       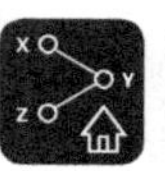  

A. What will you keep track of in the first column of your list? In the second column? In the third column? In the fourth column?

B. If you begin with a 50-gram weight, how many different combinations can you find that make 85 grams?

C. What is the greatest number of 25-gram weights you could use? How many combinations can you find using 25-gram weights with other weights?

50 grams	25 grams	15 grams	5 grams
1	1	0	2
1	0	2	1

D. Finish your list. How many different combinations of weights will balance the scale for the first liquid?

Read the problem again. Look at the information given and the main question. Review your work. Is your answer reasonable?

Make an Organized List

4

A community of daubens in the Magic Forest is upset because their forest is being bulldozed. The daubens have decided to move far away, so far that they will need to travel by boat. Each boat can hold up to 100 grams and still stay afloat. The daubens, as it happens, come in five different weights: 60 grams (senior citizens), 40 grams (adults), 20 grams (teenagers), 10 grams (children), and 5 grams (infants). How many different combinations of daubens can ride in a boat?

① FIND OUT

A. What is the question you have to answer?

B. How many different kinds of daubens are there? How much does each kind of dauben weigh?

C. How much can a boat hold safely?

② CHOOSE A STRATEGY

I can ________________________________ to solve the problem.

       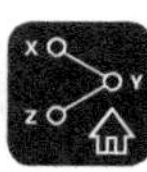  

A. What will you keep track of in the first column of your list? In the second column? The third? The fourth? The fifth?

60 grams	40 grams	20 grams	10 grams	5 grams
1	1	0	0	0
1	0	2	0	0
1	0			

B. How many grams can the boat hold? What is the maximum number of 60-gram daubens that can be in the boat? How many 40-gram daubens can be in the boat?

C. Begin with a 60-gram dauben again. How many 20-gram daubens can also be in the boat? What are all the other combinations you can find, beginning with a 60-gram dauben?

D. What is the maximum number of 40-gram daubens that could be in the boat? How many 20-gram daubens can also be in the boat? What are all the other combinations you can find, beginning with 40-gram daubens?

E. Finish your list. How many different combinations of daubens can ride in a boat?

Read the problem again. Look at the information given and the main question. Review your work. Is your answer reasonable?

Act Out or Use Objects

5

You have been hired to supervise the Fun House. The first group has 22 students. You will sit in the control booth, hidden behind four two-way mirrors. One mirror is on each wall. Mirror 1 looks into rooms A, B, and C; mirror 2 looks into rooms C, E, and H; mirror 3 looks into F, G, and H; mirror 4 looks into A, D, and F. What is a possible arrangement of the 22 students within the rooms, if you can see 9 students through each mirror?

❶ FIND OUT

A. What is the question you have to answer?

B. Where are you sitting in the Fun House? How many mirrors can you look through? Which rooms can you see through each mirror? How many different rooms are there?

C. How many students can you see through each mirror? How many students are there in all?

❷ CHOOSE A STRATEGY

I can _________________________ to solve the problem.

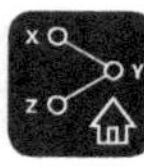

A. How many students can you see through any mirror? How many rooms are they in? What are some combinations of 3 numbers that add up to 9?

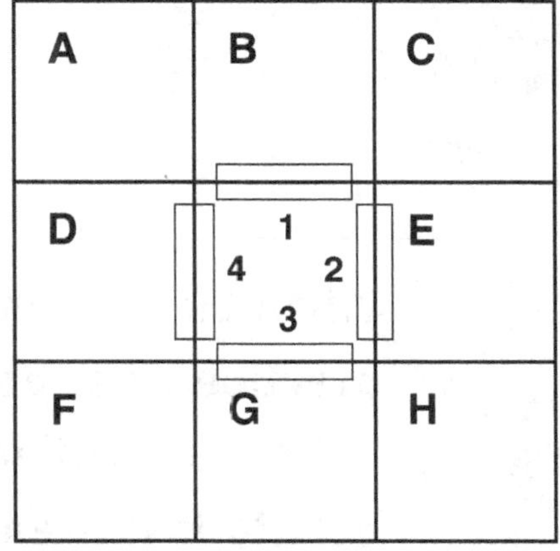

B. Would it be easier to have one scrap of paper for each of the 22 students or scraps of paper for each of the numbers that form a sum of 9? How will you label the scraps?

C. Begin with the rooms you can see through mirror 1. What do the numbers in these rooms have to add up to? What is a possible arrangement?

D. Now go to mirror 2. What number is already in room C? What can you put in E and H? Go to mirror 3. What number is in room H? What numbers can you put in F and G?

E. What number has to go in room D? Why? Do the numbers in all the rooms add up to 22?

F. Move the papers around until they match the clues.

Read the problem again. Look at the information given and the main question. Review your work. Is your answer reasonable?

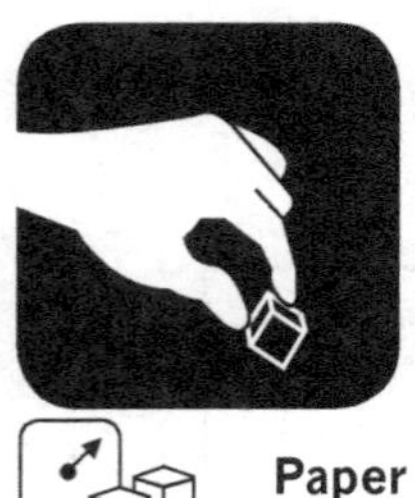

Act Out or Use Objects

6

When Liz was on her way home from hockey practice, she received several cell phone calls. Paul called before Donna. Lani called next to last. Irene called between Marco and Lani. Allen called after Dave, and Peter called before Marco. Michelle called after Lani, and Donna called before Dave. Allen called between Dave and Peter. In what order did Liz's friends call?

1 FIND OUT

A. What is the question you have to answer?

B. How many people called? What are the names of the friends who called?

C. What do you know about the order in which everyone called?

2 CHOOSE A STRATEGY

I can ______________________ to solve the problem.

       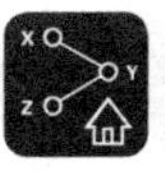  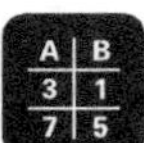

A. How many scraps of paper do you need? How will you label them?

B. Who can you put after Paul? Who goes next to last? Who called before and after Irene? Then when did Irene call? Marco? Who can go before Allen? What else do you know about Allen? Who called before Marco? Who called after Lani? Keep moving the names around until the order of callers matches the clues.

C. What was the order in which Liz's friends called?

4 LOOK BACK

Read the problem again. Look at the information given and the main question. Review your work. Is your answer reasonable?

Make an Organized List

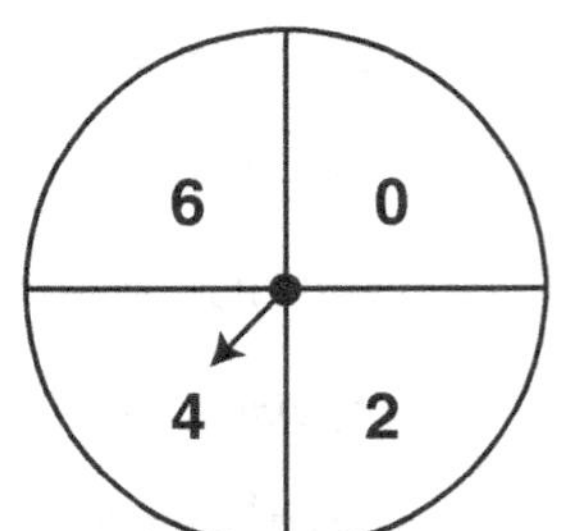

7

Kylie and Paige are playing Stock Market Mania. Kylie's game marker is 10 spaces away from the Make a Million space. She has just 3 spins left before the game ends. The spinner is divided into 4 equal parts; and the parts are labeled 0, 2, 4, and 6. The number Kylie spins tells how many spaces she may move her marker. How many different combinations of 3 spins could Kylie get to move her marker 10 spaces?

1 FIND OUT

A. What is the question you have to answer?

B. How many spaces is Kylie's game marker from Make a Million? How many spins does Kylie get before the game ends? What are the possible numbers for each spin?

2 CHOOSE A STRATEGY

I can _________________________ to solve the problem.

A. What are the possible numbers for one spin? What are the possible combinations of any 3 of those numbers that add up to 10? Finish the list of these combinations.

Combinations of 10
0 + 6 + 4

Spin 1	Spin 2	Spin 3
0	6	4
0		

B. Look at the organized list that has been started. Begin with 0 for Spin 1, 6 for Spin 2, and 4 for Spin 3. Is there another way to arrange these numbers, using 0 for Spin 1 again?

C. Now use 6 for Spin 1. How many ways can you list these same numbers, using 6 for Spin 1? Now use 4 for Spin 1. How many ways can you list the same numbers, using 4 for Spin 1?

D. Use each of the combinations. Finish your list. How many different ways could Kylie spin 3 times and get to move her marker 10 spaces?

4 LOOK BACK

Read the problem again. Look at the information given and the main question. Review your work. Is your answer reasonable?

Make an Organized List

8

It is the twins' turn to put dishes into the dishwasher and scrub the pans. Instead of fighting about who has to scrub pans, Jamal suggests a game to settle the dispute. He hands Jamar a number cube whose six faces are marked 1, 2, 3, 4, 5, and 6. Jamal explains that they have to take turns rolling the number cube 3 times. The first one to roll a total of 12 gets to put the dishes into the dishwasher tonight. How many different ways can the twins combine 3 rolls to total 12?

① FIND OUT

A. What is the question you have to answer?

B. How is the number cube marked? How many rolls does each player get? What is the winning total for 3 rolls?

② CHOOSE A STRATEGY

I can _________________________________ to solve the problem.

       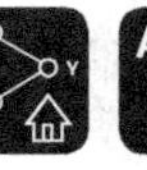  

A. What combinations of 3 numbers equal 12 points? Use only the numbers that appear on the number cube. Make a list of the combinations.

<u>**Combinations of 12**</u>
$$1 + 5 + 6$$
$$2 + 5 + 5$$

B. Look at the organized list. Begin with 1 for Roll 1, 5 for Roll 2, and 6 for Roll 3. Is there another way to arrange these numbers, using 1 for Roll 1?

Roll 1	Roll 2	Roll3

C. Now use 5 for Roll 1. How many ways can you list these same numbers, using 5 for Roll 1?

D. Now use 6 for Roll 1. How many ways can you list these same numbers, using 6 for Roll 1?

E. Work with each other set of 3 numbers that has a sum of 12. Finish making your list. How many ways can the players score 12 points?

Read the problem again. Look at the information given and the main question. Review your work. Is your answer reasonable?

Use or Look for a Pattern

9

On July 5, six different people reported to the police that they had seen Bigfoot. The next day, twice as many people called the police to report that they had seen the creature. Each day the police received twice as many calls as the day before. After they got a total of more than 300 calls, the police took the phone off the hook! On what day did the police receive their 300th call?

1 FIND OUT

A. What is the question you have to answer?

B. How many people called the police the first day? How did the number of calls change from the first day to the second day? How did the number of calls increase from day to day?

C. When did the police take the phone off the hook?

2 CHOOSE A STRATEGY

I can _________________________________ to solve the problem.

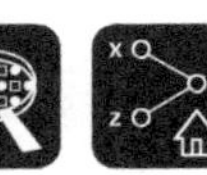

A. What is the pattern in the number of calls the police received each day?

B. You can use this table to keep track of the calls coming in each day and the total number of calls.

Day	1	2	3	
Number of Calls	6	12	24	
Total	6	18		

C. What will you keep track of in the first row of the table? In the second row? In the third row?

D. How many calls did the police get on day 1? On day 2? What was the total number of calls by the end of day 2? How many calls were there on day 3? What was the total number of calls by the end of day 3?

E. Continue filling in the table. On what day did the police receive their 300th call?

4 LOOK BACK

Read the problem again. Look at the information given and the main question. Review your work. Is your answer reasonable?

Use or Look for a Pattern

Archaeologists found a series of caves. The first cave had a circle of 560 stones. The circle in the second cave had 8 fewer stones than in the first cave. The circle in the third cave had 16 fewer stones than in the second cave. The circle in each new cave had twice as many stones missing as in the previous cave. Which cave will be the last cave with stones?

1 FIND OUT

A. What is the question you have to answer?

B. How many stones were in the circle in the first cave? What was the difference between the numbers of stones in the first and second caves? Between the second and third caves?

C. How did the number of stones change from cave to cave?

2 CHOOSE A STRATEGY

I can ________________________________ to solve the problem.

       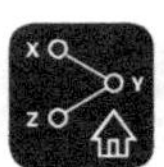  

Cave	1	2	3	4	
Fewer Stones		8	16	32	
Total in Circle	560	552			

A. What will you keep track of in the first row of the table? In the second row? In the third row?

B. What is the pattern of decrease in the stone circles?

C. How many stones are in the circle in cave 1? How many fewer stones are in the circle in cave 2? How many stones are in the circle in cave 2? How many fewer stones are in cave 3 than in cave 2? How many stones are in the circle in cave 3?

D. Continue filling in your table. Which cave will be the last cave with stones?

4 LOOK BACK

Read the problem again. Look at the information given and the main question. Review your work. Is your answer reasonable?

11

It is **12:00**, and people are lining up for the matinee. By **12:05**, there are **6** people in line. By **12:10**, there are **11** people in line. By **12:15**, there are **17** people in line. By **12:20**, there are **24** people in line, and people keep lining up in the same way all afternoon. What time will it be when there are **87** people in line?

1 FIND OUT

A. What is the question you have to answer?

B. What time do people start lining up for the matinee?

C. How many people are in line by 12:05? How many by 12:10? How many by 12:15? How many by 12:20?

2 CHOOSE A STRATEGY

I can _______________________ to solve the problem.

       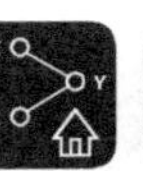  

Time	12:05	12:10	12:15	
Total in Line	6	11	17	

A. What will you keep track of in the first row of the table? In the second row? How many minutes are there between the times?

B. What is the difference between the numbers of people in line at 12:05 and 12:10? Between the numbers of people in line at 12:10 and 12:15? Between the numbers of people in line at 12:15 and 12:20?

C. What pattern do you see in those differences?

D. Use the pattern to continue filling in your table. What time will it be when 87 people are in line?

4 LOOK BACK

Read the problem again. Look at the information given and the main question. Review your work. Is your answer reasonable?

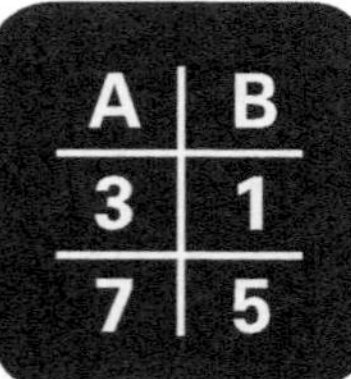

Use or Make a Table

12

Dylan made 4 Super Burgers in the first hour, 8 in the second hour, 7 in the third hour, 11 in the fourth hour, and 10 in the fifth hour. Sean made 5 Best Burgers in the first hour, 6 in the second hour, 9 in the third hour, 10 in the fourth hour, and 13 in the fifth hour. Dylan and Sean continued in the same way until they made a combined total of 58 burgers in the same hour. How many of the 58 were Super Burgers and how many were Best Burgers?

1 FIND OUT

A. What is the question you have to answer?

B. How many burgers does each boy make in each of the first 5 hours?

C. How many did they make in the last hour?

2 CHOOSE A STRATEGY

I can _________________________ to solve the problem.

Hour	1	2	3	4	5
Dylan	4	8			
Sean	5	6			
Total	9	14			

A. What will you keep track of in the first row of the table? In the second row? In the third row? In the fourth row? Fill in the numbers of burgers each boy made during hours 3, 4, and 5.

B. Look at the numbers of burgers that Dylan made. What is the difference between hours 1 and 2? Between hours 2 and 3? Between hours 3 and 4? Between hours 4 and 5? What pattern do you see in the numbers of burgers Dylan made?

C. Look at the numbers of burgers that Sean made. What is the difference between hours 1 and 2? Between hours 2 and 3? Between hours 3 and 4? Between hours 4 and 5? What pattern do you see in the numbers of burgers Sean made?

D. Use the patterns to continue filling in the table. When did they make a combined total of 58? How many of the 58 burgers were Supers, and how many were Bests?

Read the problem again. Look at the information given and the main question. Review your work. Is your answer reasonable?

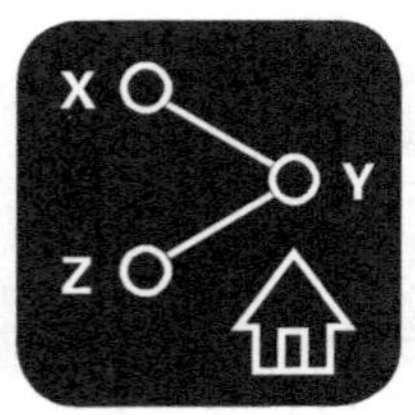

13

Venski and his brother have each agreed to mow half of the lawn. The lawn is a rectangle measuring 120 feet by 60 feet. The lawn mower cuts a 4-foot-wide strip. Venski starts at one corner and mows around the edges of the lawn. He comes closer to the center with each trip. On which trip around will he have half the area of the lawn mowed?

1 FIND OUT

A. What is the question you have to answer?

B. What are the shape and dimensions of the lawn? How wide is the strip of lawn that the mower cuts as it moves?

C. Where does Venski start mowing the lawn? How does he move as he mows it?

2 CHOOSE A STRATEGY

I can _________________________________ to solve the problem.

A. What is the length of the lawn? What is the width of the lawn? What is the area of the lawn? What is half the area?

B. How wide is the strip cut by the mower? Look at how the first strip is shown on the diagram.

C. After the first strip has been mowed, what are the dimensions of the part still to be mowed? What is the area of the part still to be mowed? Is it more than half, or less? Continue to show the results of each trip on your diagram until the area of the part to be mowed is half or less than half of the lawn's area.

D. On which trip will Venski have mowed his half?

4 LOOK BACK

Read the problem again. Look at the information given and the main question. Review your work. Is your answer reasonable?

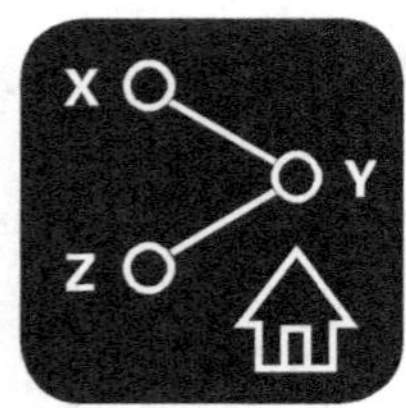

Use or Make a Picture or Diagram

14

Holly and Hal are working in a garden with a statue at the center. A series of rectangular borders with a different type of flower in each surround the statue. The statue's base is 12 feet by 2 feet. Each border is a strip 3 feet wide. The daisy border has 8 times the area of the statue's base. What are the dimensions of the daisy border?

1 FIND OUT

A. What is the question you have to answer?

B. What is the shape of the base of the statue? What are the dimensions of the base? How wide is each border surrounding the statue?

C. What do you know about the area of the daisy border?

2 CHOOSE A STRATEGY

I can ______________________________ to solve the problem.

       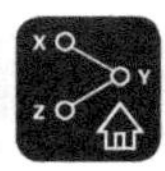  

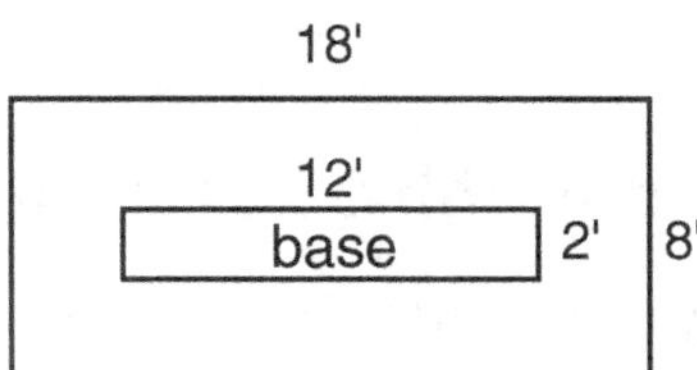

A. What are the dimensions of the base? What is the area of the base? What is 8 times that area?

B. How wide is each border? The first border is shown on the diagram. Show the second border. Write the dimensions on the outer edges.

C. What is the area of the base plus the first border? What is the area of the first border alone? Is it 8 times the area of the base? Continue drawing borders and labeling the dimensions until you draw the border with the right area.

D. What are the dimensions of the daisy border?

④ **LOOK BACK**

Read the problem again. Look at the information given and the main question. Review your work. Is your answer reasonable?

Guess and Check

15

How many minutes did Abby, Jesse, and Joy each travel to get to the skating rink on Saturday? Joy came by skateboard, Abby came by bike, and Jesse came on the bus. It took Abby twice as long as Joy to get there. It took Jesse 10 minutes more than the time it took both the girls together. The total time it took all three skaters to get to the rink adds up to 64 minutes.

❶ FIND OUT

A. What is the question you have to answer?

B. What do you know about Abby's traveling time? Jesse's traveling time? Joy's traveling time?

C. How long did it take all three skaters together to get to the rink?

❷ CHOOSE A STRATEGY

I can ________________________ to solve the problem.

A. Will you start by making a guess for Abby, Jesse, or Joy? Why? What is your guess?

B. If you make a guess for one person, how can you figure out the time taken by each of the other two?

C. How can you check your guess? How was your guess? If your first guess was wrong, how can you make your next guess better?

D. How many minutes did Abby, Jesse, and Joy each travel to get to the skating rink on Saturday?

4 LOOK BACK

Read the problem again. Look at the information given and the main question. Review your work. Is your answer reasonable?

Guess and Check

16

Auditions are next week. Tamara wants a part that has twice as many lines as the part that Berta wants. Adam has chosen a part with three lines more than twice as many as the part that Tamara wants. Renee is going to try out for a part with four more lines than Berta's part. If they get the parts they want, together they will have a total of 47 speaking lines. How many lines would each actor have in the play?

1 FIND OUT

A. What is the question you have to answer?

B. What do you know about Tamara's part? Adam's part? Renee's part? Berta's part?

C. What is the total number of speaking lines for the four actors?

2 CHOOSE A STRATEGY

I can _________________________________ to solve the problem.

A. Will you begin with a guess for Tamara, Adam, Renee, or Berta? Why? What is your guess?

B. If you make a guess for one of the actors, how can you figure out the number of lines for the other actors?

C. How can you check your guess? How was your guess? If your first guess was wrong, how can you make your next guess better?

D. How many lines would each actor have in the play?

4 LOOK BACK

Read the problem again. Look at the information given and the main question. Review your work. Is your answer reasonable?

Work Backwards

> **Kevin, Barbara, and their parents went backpacking. On the first and second days, each hiker ate a serving of food for breakfast, for lunch, and for dinner. A large brown bear barged into camp the second night, got the food pack down from the tree, and ate one-half of the food that was left. The next morning, after they all had breakfast, they had 4 food servings left. How many servings of food did they begin the trip with?**

1 FIND OUT

A. What is the question you have to answer?

B. How many hikers were there? What do you know about the servings eaten on day 1 and 2?

C. How much did the bear eat? How many servings were left after the bear finished eating?

2 CHOOSE A STRATEGY

I can _________________________________ to solve the problem.

Look at the list that has been started.

Day 3 - 4 left
4 eaten

A. Begin with the last day of the trip. How many servings of food did they have left after breakfast?

Day 2 -

B. How many servings did the hikers eat on the last day? How many servings did the hikers have all together before breakfast on the last day?

C. How can you figure out how much the bear ate?

D. How many servings were eaten on the second day? How many servings were eaten on the first day? How many servings did they begin with?

4 LOOK BACK

Read the problem again. Look at the information given and the main question. Review your work. Is your answer reasonable?

Work Backwards

18

Out of his usual morning batch of chocolate chip cookies, Bill burned the first two dozen. He gave $\frac{1}{2}$ of what was left to Keri to take to school. Next, he wrapped up $\frac{1}{2}$ of the remaining cookies and gave them to the firehouse crew. Then he gave $\frac{1}{2}$ of what was left to a police officer. If Bill had only 7 cookies left, how many cookies were in the batch ?

1 FIND OUT

A. What is the question you have to answer?

B. What happened to Bill's cookies first? How many cookies did Bill give Keri? The firehouse crew? The police officer?

C. How many cookies did Bill have left?

2 CHOOSE A STRATEGY

I can _________________________ to solve the problem.

       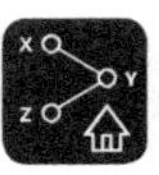  

A. Look at the diagram that has been started. How is the diagram divided? What equal parts do you see?

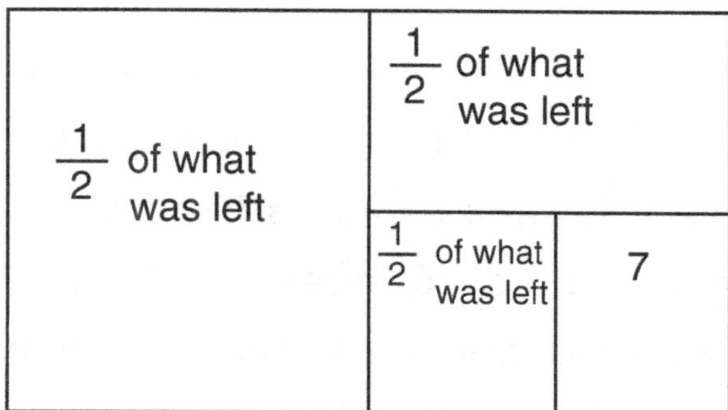

B. Begin with what Bill had left after he burned and gave away cookies: 7 cookies. Who is the last person Bill gave cookies to? How many?

C. Continue to work backwards. Who did he give cookies to before that? How many?

D. Who is the other person Bill gave cookies to? How many?

E. What is the first thing that happened to the cookies?

F. How many cookies were in the batch?

Read the problem again. Look at the information given and the main question. Review your work. Is your answer reasonable?

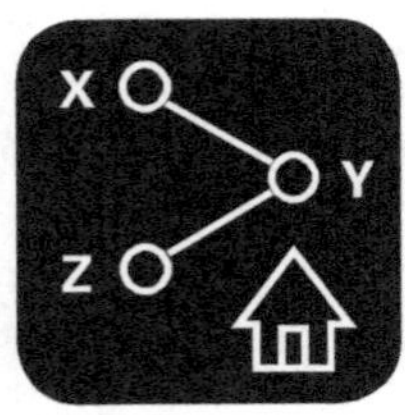

Use or Make a Picture or Diagram

19

When Larry and Brett's car broke down, the brothers decided to take turns on Brett's skateboard to travel the 36 miles into town. They want to leave at the same time and arrive in town at the same time. Each boy can skateboard 6 miles per hour. Larry can walk 3 miles an hour and Brett can walk 4 miles an hour. Each boy can only leave the skateboard for his brother on the hour, but neither one can wait for the other. How long will it take them to get to town, and how many miles will each one walk?

① FIND OUT

A. What is the question you have to answer?

B. How will they travel? At what speeds? When do they want to arrive in town? What are the rules for exchanging the skateboard?

② CHOOSE A STRATEGY

I can _________________________ to solve the problem.

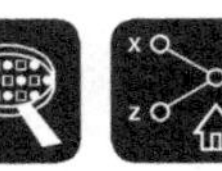

Look at the diagram that has been started.

Hour	1	2	3	4
Larry	6 sk	12 sk		
Brett	4 wk	8 wk	12 wk	

A. What will you keep track of in your diagram?

B. If Larry is the slower walker, then maybe he could start on the skateboard. After one hour, where will each one be?

C. If they can only exchange the skateboard on the hour, when does Larry need to leave the skateboard so that Brett can pick it up? When will the next skateboard exchange be? Complete your diagram.

D. How long will it take the brothers to get to town and how many miles will each one walk?

4 LOOK BACK

Read the problem again. Look at the information given and the main question. Review your work. Is your answer reasonable?

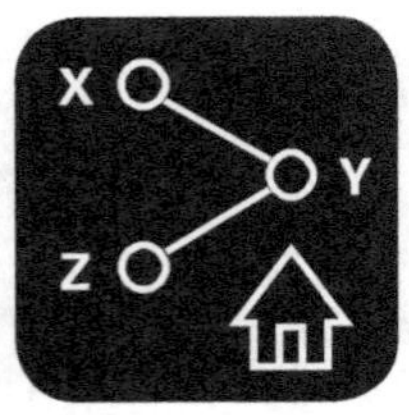

Use or Make a Picture or Diagram

20

James, Matt, and Kate each traveled at different speeds on the same road in Spain. James decided to hike 12 km the first day, rest the second day, rent a bike on the third day to travel 21 km, and hike 12 km to Burgos on the fourth day. Matt will rest the first day, bike 19 km the second day, hike 14 km the third day, and hike 12 km the fourth day. Kate plans on hiking all four days and wants to spend as much time as possible hiking with each of her friends. How should Kate plan her travel, and with whom will she travel each day?

1 FIND OUT

A. What is the question you have to answer?

B. How long will the trip last? How will the friends travel?

C. What is Kate's goal for her hiking itinerary?

2 CHOOSE A STRATEGY

I can _________________________________ to solve the problem.

       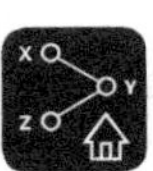  

	James	Matt	Kate
Day 1	+ 12 km, hiking = 12 km	+ 0 km, resting = 0 km	
Day 2			
Day 3			
Day 4			

A. What do we know about James' plans? What do we know about Matt's plans?

B. What will you keep track of in each column of the diagram? What will you keep track of in each row?

C. How far does James travel on Day 1? What about Matt? Who can Kate hike with and how far?

D. How far does James travel on Day 2? What about Matt? If Kate wants to start Day 3 in the same place as one of her friends, how far will she go on Day 2?

E. How can we fill in the chart for Day 3? What about Day 4?

F. How should Kate plan her travel, and with whom will she travel each day?

4 LOOK BACK

Read the problem again. Look at the information given and the main question. Review your work. Is your answer reasonable?

Use Logical Reasoning

21

The square and circle on this ancient stone tablet represent the number 14.

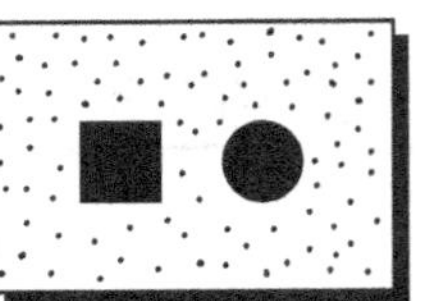

The squares and circles on the second tablet represent the number 34.

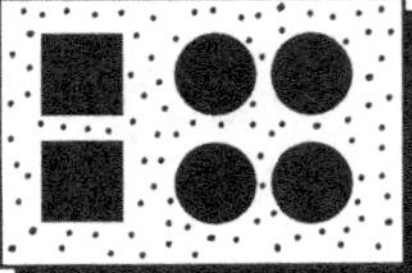

What number do the square and circles on the third tablet represent?

---

1 FIND OUT

A. What is the question you have to answer?

B. Find out what the problem tells you.

---

2 CHOOSE A STRATEGY

I can _________________________ to solve the problem.

        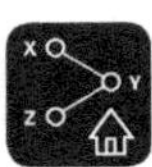  

A. What do you know about the first tablet?

B. Try guessing some numbers for the square and circle. Which numbers will add up to 14?

C. If you find two numbers that work for the first tablet, then what is the next step? When you have values for the square and circle that give you 14 and 34 for the first and second tablets, how can you find a value for the mystery tablet?

D. What number do the square and circles on the third tablet represent?

4 **LOOK BACK**

Read the problem again. Look at the information given and the main question. Review your work. Is your answer reasonable?

Use Logical Reasoning

22

Gnomes from all over the forest are eager to take a hot-air balloon flight. The balloon will only support the weight of 24 gnome babies. The weight of 4 children equals the weight of 8 babies, and the weight of 6 children is equal to the weight of 3 adults. How many adult gnomes or gnome children will the balloon support?

① FIND OUT

A. What is the question you have to answer?

B. How much weight will the balloon support?

C. What is the weight of 4 gnome children equal to? 6 gnome children?

② CHOOSE A STRATEGY

I can _________________________ to solve the problem.

       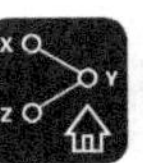  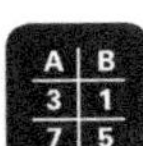

A. You can begin by writing as equations the information given in the problem.

If
4 children =
6 children =
Then

B. If the weight of 4 gnome children equals the weight of 8 babies, then how many children together would equal the weight of 24 babies?

C. If the weight of 6 children equals the weight of 3 adults, then how many adults together would equal the weight of 12 children?

D. If that is so, then how many adults together would equal the weight of 24 babies?

E. How many adult gnomes will the balloon support, and how many gnome children will the balloon support?

4 LOOK BACK

Read the problem again. Look at the information given and the main question. Review your work. Is your answer reasonable?

Use or Make a Table

23

Travis works at Fantasy in Flight Factory. He checks all the kites before they are packaged. Travis discovered that for every 30 kites that passed inspection, there were 7 kites that didn't pass: 4 kites did not have tails, and 3 kites had the wrong colors. Of the 296 kites Travis examined, how many didn't have tails, and how many had the wrong colors?

① FIND OUT

A. What is the question you have to answer?

B. For every 30 kites that passed inspection, how many kites did not pass inspection? How many of those didn't have tails? How many had the wrong colors?

C. How many kites did Travis examine?

② CHOOSE A STRATEGY

I can ________________________________ to solve the problem.

Pass	30	60	90	
No Tail	4	8		
Wrong Color	3	6		
Total	37	74		

A. What will you keep track of in the first row of the table? In the second row? In the third row? In the fourth row? Why do you need to keep track of the total?

B. How many kites didn't pass inspection for 90 kites that did? How many were missing a tail? How many had the wrong colors? Continue to fill in the table. How will you know when your table is complete?

C. Of the 296 kites Travis examined, how many didn't have tails, and how many had the wrong colors?

4 LOOK BACK

Read the problem again. Look at the information given and the main question. Review your work. Is your answer reasonable?

Use or Make a Table

24

The Rialto Theater is celebrating its 11th anniversary by giving away free passes! They have hidden a gold, silver, purple, or green star under every seat. Every person who sits in a seat that has a gold star gets a free pass to the next show. For every 2 gold stars, they hid 18 silver, 16 purple, and 12 green stars. If there are 384 seats in the Rialto Theater, how many people will win free passes?

1 FIND OUT

A. What is the question you have to answer?

B. For every 2 gold stars, how many silver were there? How many purple? How many green?

C. How many seats are there in the theater?

2 CHOOSE A STRATEGY

I can _______________________________ to solve the problem.

Gold	2	4	
Silver	18	36	
Purple	16	32	
Green	12		
Total	48		

A. What will you keep track of in the first row of the table? In the second row? In the third row? In the fourth row? In the fifth row? Why do you need to keep track of the total number of stars?

B. If there are 2 gold stars, how many seats in all have stars on them? If there are 4 gold stars, how many silver stars are there? How many purple? How many green? How many seats in all have stars on them?

C. Continue to fill in the table. When can you stop?

D. How many people will win free passes?

4 LOOK BACK

Read the problem again. Look at the information given and the main question. Review your work. Is your answer reasonable?

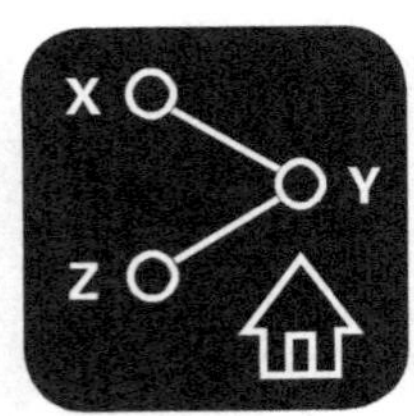

Use or Make a Picture or Diagram

25

Each team in the canoe relay race must have two adults and three campers. In order to win, all of the members of a team must canoe from one shore to the other shore. Teams must follow these rules for who can be in the canoe on a crossing: one adult; one camper; one adult and one camper; or two campers. What is the smallest number of trips a team would have to make to win the relay?

➊ FIND OUT

A. What is the question you have to answer?

B. How are the teams organized for the race?

C. How does a team win the race? What are the rules for each river crossing?

➋ CHOOSE A STRATEGY

I can _________________________________ to solve the problem.

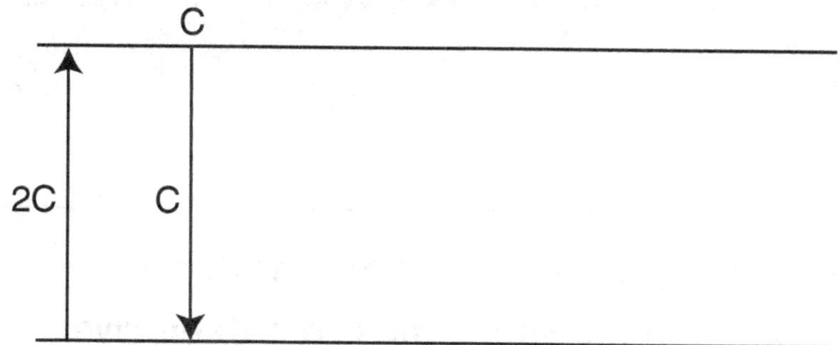

A. What do you want to keep track of in your diagram?

B. Begin with two campers crossing the river. What is the next step? How many team members have you gotten across?

C. If one camper returns on crossing 2, what are your choices for crossing 3? Keep adding trips to the diagram.

D. What is the smallest number of trips a team would have to make to win the relay?

4 LOOK BACK

Read the problem again. Look at the information given and the main question. Review your work. Is your answer reasonable?

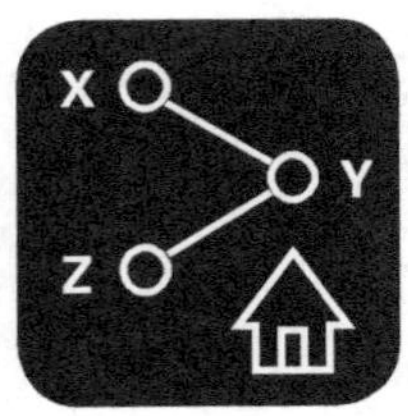

26

Jessica is hiking in the mountains with her llama, dog, and cat. She comes to a deep river where she must help each animal across, one at a time. She has a problem: she can't leave the llama alone with the dog, and she can't leave the cat alone with the dog. What is the smallest number of trips Jessica can make to get the llama, the dog, and the cat to the other side of the river?

1 FIND OUT

A. What is the question you have to answer?

B. What are the conditions for crossing the river? What are the special conditions for the llama and the dog? What are the special conditions for the cat and the dog?

2 CHOOSE A STRATEGY

I can _______________________ to solve the problem.

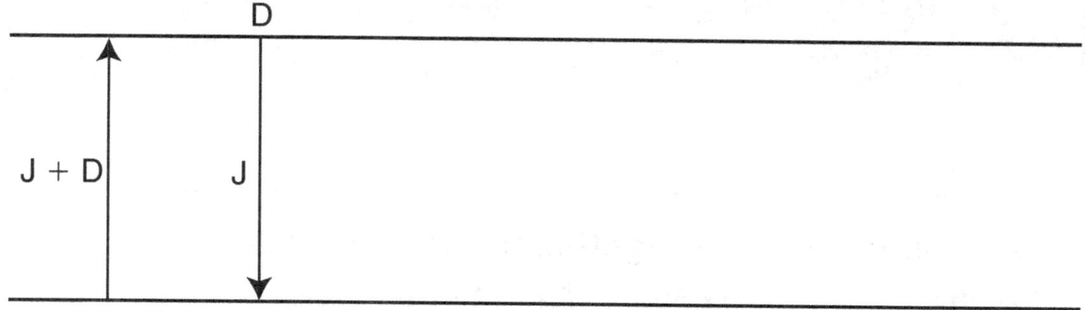

A. What do you need to keep track of in your diagram?

B. Who must make the first crossing? Why?

C. If Jessica returns alone on crossing 2, who would be a good choice for crossing 3? Keep showing crossings on the diagram.

D. What is the smallest number of trips Jessica can take to get everyone across?

4 LOOK BACK

Read the problem again. Look at the information given and the main question. Review your work. Is your answer reasonable?

Make an Organized List

27

Laura is in charge of lighting the Rock Palace for the upcoming concert. Each light fixture supplies exactly 1,000 watts of power to light the bulbs in the fixture. Laura can use any combination of 150-watt, 100-watt, 75-watt, or 60-watt bulbs, but the total number of watts must be 1,000. How many different combinations of bulbs could Laura use in a light fixture?

1 FIND OUT

A. What is the question you have to answer?

B. What must the total number of watts be for each combination of bulbs?

C. What sizes of light bulbs can Laura use?

2 CHOOSE A STRATEGY

I can _________________________ to solve the problem.

       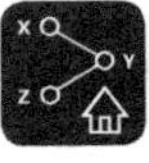  

A. What will you keep track of in the first column of your list? In the second column? In the third column? In the fourth column?

150W	100W	75W	60W
6	1	0	0
5	1	2	0
4	4	0	0
4	1	4	0

B. What is the greatest number of 150-watt bulbs Laura can use? What is their total wattage? Does Laura need to use any 100-watt bulbs with the 150-watt bulbs? How many? Any 75-watt bulbs? Any 60-watt bulbs? Is there any other combination that could be made with six 150-watt bulbs?

C. Is there a combination that could be made with five 150-watt bulbs? What is their total wattage? Does Laura need to use any 100-watt bulbs? How many? Any 75-watt bulbs? How many? Is there any other combination that could be made with five 150-watt bulbs? Continue filling in your list.

D. How many different combinations of bulbs could Laura use in a light fixture?

Read the problem again. Look at the information given and the main question. Review your work. Is your answer reasonable?

Make an Organized List

Arturo is selling ads for his school paper, *The Gab Gazette.* He has only one page left to fill. The sizes of his ads are $\frac{1}{2}$, $\frac{1}{4}$, $\frac{1}{8}$ and $\frac{1}{16}$ of a page. Before Arturo sells more ads, he has to figure out all of the different combinations of ads that would fill one page. How many different combinations would fill one page?

28

1 FIND OUT

A. What is the question you have to answer?

B. How many different sizes of ads does Arturo have? What are the sizes?

C. How much space does Arturo have left to fill?

2 CHOOSE A STRATEGY

I can _________________________ to solve the problem.

A. What will you keep track of in the first column? Second column? Third column? Fourth column?

$\frac{1}{2}$ Page	$\frac{1}{4}$ Page	$\frac{1}{8}$ Page	$\frac{1}{16}$ Page
2	0	0	0
1	2	0	0
1	1	2	0
1	1	1	2

B. What is the largest number of $\frac{1}{2}$-page ads you could use? What is the total space covered by those ads? Do you have to use any $\frac{1}{4}$-page ads, $\frac{1}{8}$-page ads, or $\frac{1}{16}$-page ads with the $\frac{1}{2}$-page ads?

C. Find all the different combinations possible, using one $\frac{1}{2}$-page ad. How many different combinations did you make? Continue filling in your list in an organized way so that you will find every possible combination.

D. How many different combinations of ads would fill one page?

Read the problem again. Look at the information given and the main question. Review your work. Is your answer reasonable?

Guess and Check

29

Raul and Derrick are playing a game called Ski Patrol. They are on the ski patrol and have to make different runs down the mountain, sometimes making ski jumps along the way. They get 15 points for a ski run completed, and 20 points for a ski jump completed successfully. Together Raul and Derrick score 430 points for making a combined total of 25 ski runs and jumps. How many ski runs and how many ski jumps have they made?

1 FIND OUT

A. What is the question you have to answer?

B. How are points scored in Ski Patrol?

C. How many points do Raul and Derrick have? What are their points for?

2 CHOOSE A STRATEGY

I can _________________________________ to solve the problem.

         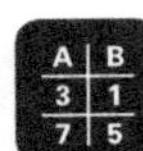

A. What is the total number of ski runs and ski jumps that they made? How many points did they get for a ski run? For a ski jump?

B. Make a guess, keeping in mind that the total number of runs and jumps must be 25. What is your guess for ski runs? For ski jumps?

C. How can you check your guess? How did you do? If your guess was wrong, how can you use that information to make your next guess?

D. Keep making guesses and checking them until you get the right answer. How many ski runs and how many ski jumps have they made?

4 LOOK BACK

Read the problem again. Look at the information given and the main question. Review your work. Is your answer reasonable?

Guess and Check

30

At a recent Insect Convention of manypeds and multipeds held at the Hotel Fleeflea, the desk clerk registered 293 guests. He noticed that each manyped had 16 legs and that each multiped had 14 legs. Seth, the shoeshine spider, reported that there were 4,408 legs in total at the convention. How many of the registered guests were manypeds, and how many were multipeds?

❶ FIND OUT

A. What is the question you have to answer?

B. What did the desk clerk notice?

C. What did Seth report to the desk clerk?

D. How many guests were registered?

❷ CHOOSE A STRATEGY

I can _________________________ to solve the problem.

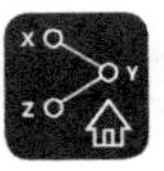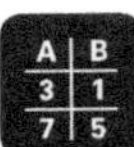

A. What was the total number of multipeds and manypeds?

B. Make a guess, and remember that the total number of guests must be 293. How many multipeds? How many manypeds?

C. How many legs did each multiped have? Each manyped?

D. How can you check your guess? How did you do? If your guess was wrong, how can you use that information to make your next guess?

E. Keep making guesses and checking them until you get the right combination. How many of the registered guests were manypeds, and how many were multipeds?

Read the problem again. Look at the information given and the main question. Review your work. Is your answer reasonable?

Use Logical Reasoning

31

The Yamada family has one grandmother and one grandfather, five mothers and five fathers. Each mother has three daughters, and each daughter has two brothers. The grandmother has only one daughter-in-law. What is the smallest number of family members there could be?

1 FIND OUT

A. What is the question you have to answer?

B. How many grandparents are in the family? How many parents?

C. How many daughters does each mother have? How many brothers does each daughter have?

D. How many daughters-in-law does the grandmother have?

2 CHOOSE A STRATEGY

I can _________________________________ to solve the problem.

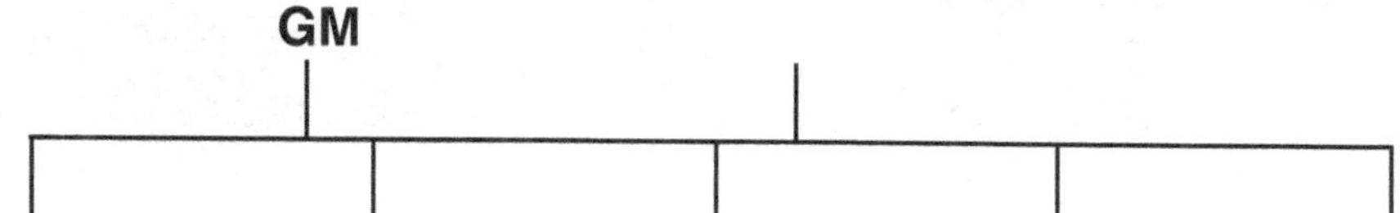

A. Is it possible for one person to have several family titles?

B. Who goes at the top of the family tree? How many daughters do they have? If each daughter has 2 brothers, then what is the smallest number of sons the grandmother can have? Where will you put these children?

C. What other relative does the grandmother have? Where will we put her?

D. Which of the women in your diagram must be mothers? How many children does each mother have? Where will you put those children?

E. How many fathers must there be? Which fathers are already shown? Where can you put the other fathers?

F. What is the smallest number of family members there could be?

Read the problem again. Look at the information given and the main question. Review your work. Is your answer reasonable?

Use Logical Reasoning

32

Molly's Music and Movies was having a one-day sale on CDs and DVDs. There were 165 people who bought CDs and 195 people who bought DVDs. 125 customers bought ONLY CDs. Some people bought both CDs and DVDs, and 60 customers did not buy CDs or DVDs. How many customers were at Molly's sale?

1 FIND OUT

A. What is the question you have to answer?

B. How many people bought DVDs?

C. How many people bought CDs? How many of those people bought ONLY CDs?

D. How many people did not buy either sale item?

2 CHOOSE A STRATEGY

I can _________________________ to solve the problem.

A. What does the circle on the left stand for? The circle on the right? Then what does section B, which is in both circles, stand for?

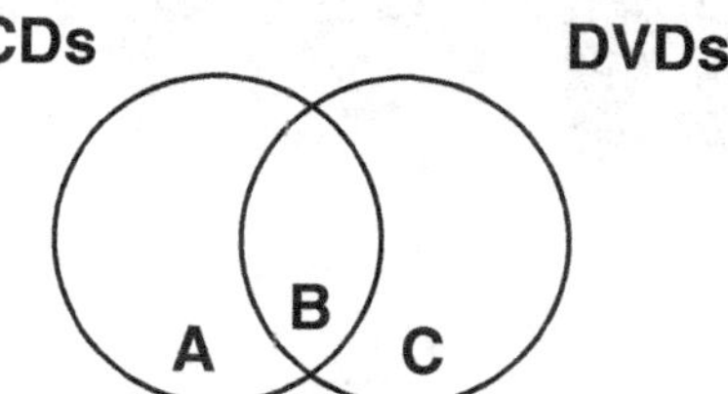

B. Start with CDs. In all, how many customers bought CDs? How many of those people bought ONLY CDs? Where do you put this number? What must the total number in the CDs' circle be? If this circle must have that total, then how many customers belong in the other part of this circle?

C. Now look at DVDs. In all, how many customers bought DVDs? The problem doesn't give more information about DVDs. Look at the diagram. If we know the number in part B, then how can we find the number in part C? How many people bought ONLY DVDs?

D. Were there customers who don't belong in either circle? Where does this number go?

E. How can you find the total number of customers?

Read the problem again. Look at the information given and the main question. Review your work. Is your answer reasonable?

Use or Look for a Pattern

33

In the first stage, the grate looked like this:

In the second stage, the grate looked like this:

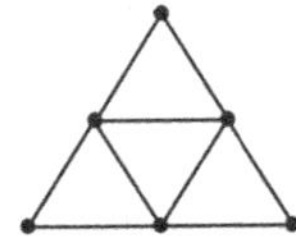

In the third stage, the grate looked like this:

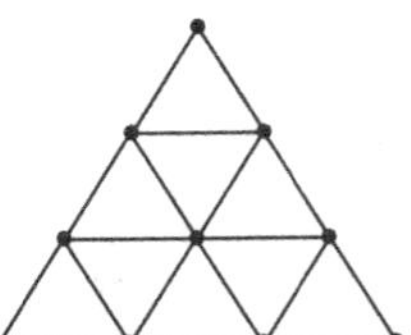

The builders kept using joints and 1-foot-long pipes to enlarge the grate in the same way until it had a 24-foot perimeter. How many joints did the builders use?

1 FIND OUT

A. What is the question you have to answer?

B. How many parts were in the grate at each stage? How long was each pipe? What was the perimeter of the finished grate?

2 CHOOSE A STRATEGY

I can _________________________ to solve the problem.

A. How many joints were in the grate at stage 1? What was the grate's perimeter?

B. How many joints were in the grate at stage 2? What was the grate's perimeter?

C. How many joints were in the grate at stage 3? What was the grate's perimeter?

D. How did the number of joints change from stage 1 to stage 2? From stage 2 to stage 3? How will the number of joints change from stage 3 to stage 4? What is the pattern of change?

Stage	Joints	Perimeter in Feet
1	3	3
2	6	6
3	10	9
4		
5		
6		
7		
8		

E. How did the perimeter change from stage 1 to stage 2? From stage 2 to stage 3? How will it change from stage 3 to stage 4? What is the pattern?

F. Continue filling in the table until the perimeter is 24 feet. How many joints did the builders use?

4 LOOK BACK

Read the problem again. Look at the information given and the main question. Review your work. Is your answer reasonable?

Use or Look for a Pattern

34

The students' snowball pyramid had a square base. The top was a single snowball.

The pictures show the top 2 layers, the top 3 layers, and then the top 4 layers.

The pyramid had 10 layers, and the sizes of the layers increased in the same way. How many snowballs did the students use?

① FIND OUT

A. What is the question you have to answer?

B. What shape was the base of the pyramid? How many layers did the pyramid have?

C. How many snowballs were in the top layer of the pyramid? The top two layers? The top three layers? Each other layer?

② CHOOSE A STRATEGY

I can _________________________________ to solve the problem.

Layers	Snowballs	Change
1	1	
2	5	
3	14	
4		
5		
6		
7		
8		
9		
10		

A. What will you keep track of in the first column of the table? In the second column? In the third column?

B. How did the number of snowballs change when layer 2 was added? Layer 3? Fill in the number of snowballs in layer 4. What is the change?

C. What is the pattern of change?

D. Continue filling in the table through the 10th layer. How many snowballs did the students use?

4 LOOK BACK

Read the problem again. Look at the information given and the main question. Review your work. Is your answer reasonable?

Use Logical Reasoning

35

Mr. Bolt, Mr. Cruz, Mr. Wells, Mrs. Gorka, and Mrs. Rojo each brought his or her child to Parent Night. The children are Ramiro, Mara, Joel, Ken, and Julie. Mr. Cruz's daughter did not inherit his freckles; Mr. Bolt's son looks just like him; Mrs. Gorka and her child have dimples; Mara and Ken are the only students with freckles, and also the only students with dimples; and Ken's father and Ramiro's father were unable to attend. Can you match each student with a parent?

1 FIND OUT

A. What is the question you have to answer?

B. Who are the parents? Who are the students?

C. What do you know about Mr. Cruz? Mr. Bolt? Mrs. Gorka? Mara and Ken? Ken and Ramiro?

2 CHOOSE A STRATEGY

I can ________________________________ to solve the problem.

	Mara	Ken	Ramiro	Joel	Julie
Mr. Bolt					
Mr. Cruz					
Mr. Wells					
Mrs. Gorka					
Mrs. Rojo					

A. What do you know about Mr. Cruz? If he has a daughter, then which students are not Mr. Cruz's child? Look at the rest of the clues. Which of the girls can't be Mr. Cruz's daughter? Then who is Mr. Cruz's daughter? If that is so, can she match up with another parent? Then where can you write *Yes* and *No*?

B. What do you know about Mr. Bolt? Then which students can't be Mr. Bolt's child?

C. Who can be Mrs. Gorka's child? Who can't be?

D. Where can you write *No* for Ken and Ramiro?

E. In any row or column, is there just one place left to write *Yes*? Then where can you write *No*?

4 LOOK BACK

Read the problem again. Look at the information given and the main question. Review your work. Is your answer reasonable?

Use Logical Reasoning

36

Aaron, Chloe, Jake, and Grace each hold a different one of these jobs: nanny, mail carrier, police officer, and tutor. Aaron and Grace work with children. Chloe and Jake must wear uniforms. Jake doesn't wear a badge. Grace works at a desk. The 20-year-old delivers letters. The nanny is 5 years older than the mail carrier. Chloe, 24 years old, is 3 years younger than the tutor. Can you match each person with an age and a job?

1 FIND OUT

A. What is the question you have to answer?

B. Who are the people? What are their jobs? What are the clues about each person's job?

C. What do you know about the people's ages?

2 CHOOSE A STRATEGY

I can _______________________ to solve the problem.

	Nanny	Police Officer	Tutor	Mail Carrier	20	24		
Aaron								
Jake								
Chloe								
Grace								

A. What will you keep track of in the rows? In the columns? What information is missing in the labels on the columns?

B. If Aaron and Grace work with children, then which jobs do they have? Where can you write *No*? If Grace works at a desk, then where else can you write *No* for Grace? Then which job does Grace have? Now where can you write *No*?

C. What do you know about Grace's age? Then what label can you put on one of the blank columns? Use the clue about Grace's age. In which 2 spaces can you write *Yes*? Then where can you write *No*?

D. Keep using the clues. Can you match each person with an age and a job?

4 LOOK BACK

Read the problem again. Look at the information given and the main question. Review your work. Is your answer reasonable?

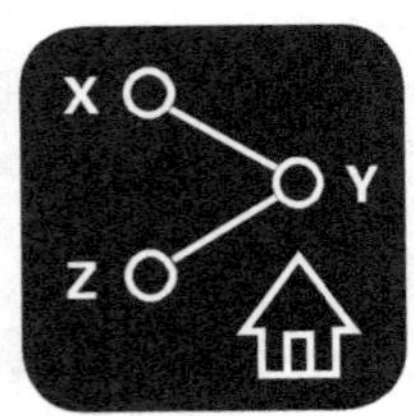

Use or Make a Picture or Diagram

37

The graph shows Brian, Ava, and Chris' sales of games, comic books, and action figures. For 6 months, fewer comic books were sold than the other items. In one three-month period Brian sold 40 items. During the summer, games sold better than comic books or action figures, and Ava's sales were lower than her friends'. What is each friend selling on the Internet?

1 FIND OUT

A. What is the question you have to answer?

B. What do you know about the comic book sales? Game sales? Action figure sales? What do you know about Brian? Ava? Chris?

2 CHOOSE A STRATEGY

I can _________________________ to solve the problem.

       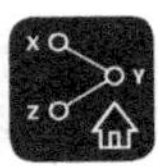  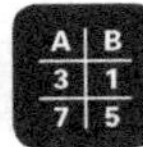

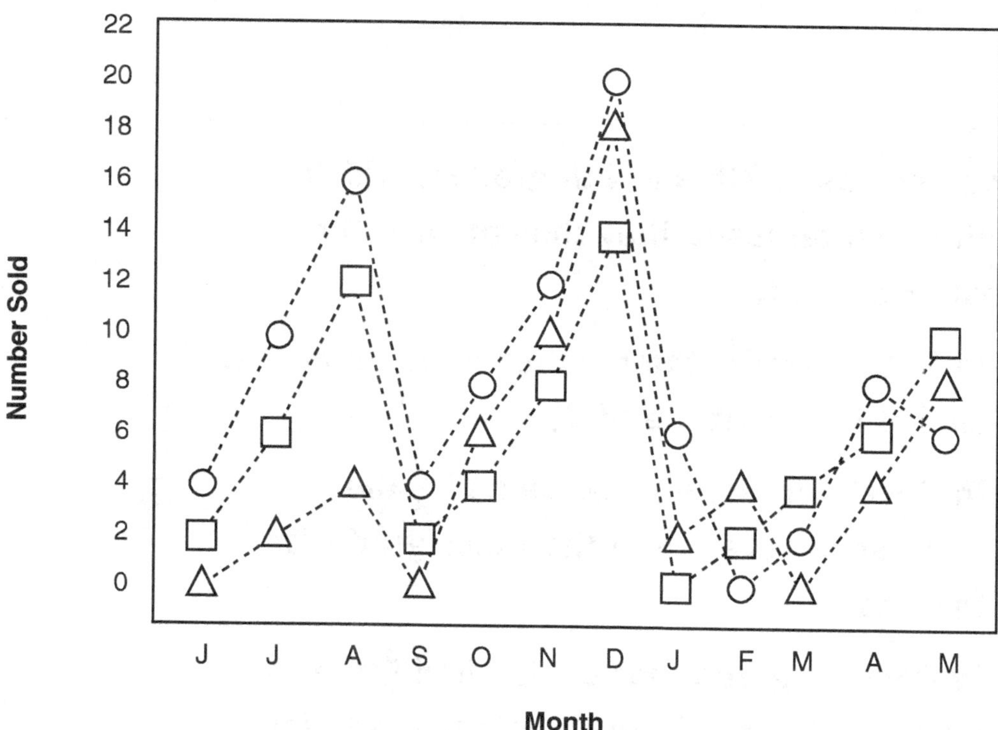

A. What will you look for in the graph to find the comics line?

B. How will you find the item that Brian sold?

C. How will you find the games line? Then which line shows action figures?

D. Use the last clue. What is each friend selling?

4 LOOK BACK

Read the problem again. Look at the information given and the main question. Review your work. Is your answer reasonable?

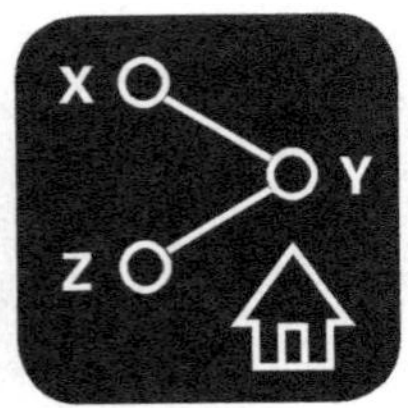

Use or Make a Picture or Diagram

38

A survey asked 7th and 8th graders which environmental issue they thought was the most important.

- In Ruth's grade, the number of votes for her choice is a multiple of 3, 6, and 8.

- In Dave's grade, his choice had $\frac{2}{3}$ the number of votes from the other grade for that issue.

- In Peggy's grade, her choice had $\frac{5}{8}$ the number of votes from the other grade for that issue.

Which grade is each student in? Which issue did each student choose?

1 FIND OUT

A. What are the questions you have to answer?

B. Find out what the problem tells you.

2 CHOOSE A STRATEGY

I can _______________________________ to solve the problem.

      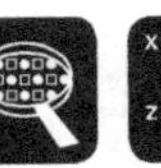 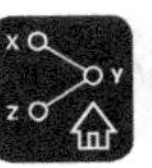  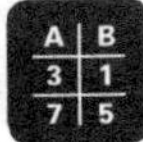

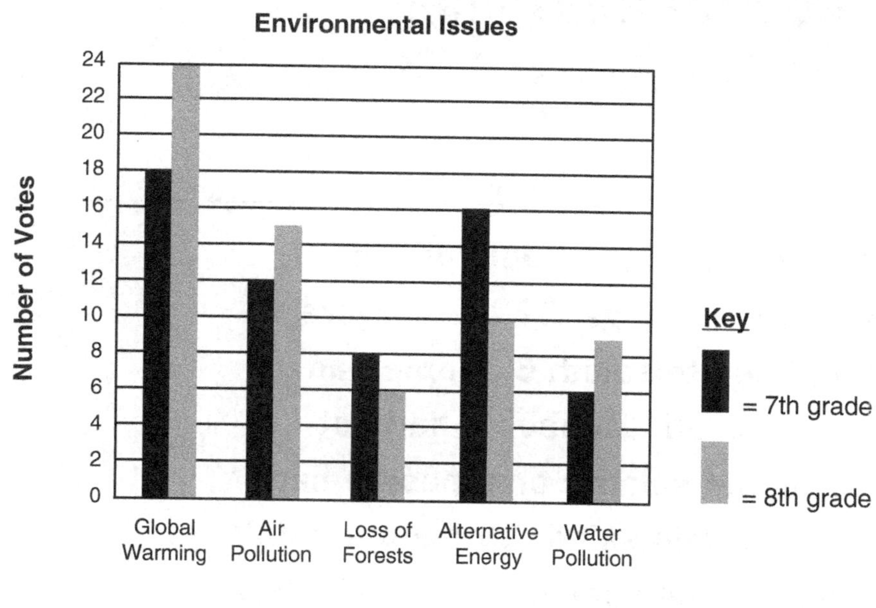

A. What is shown on the graph's horizontal axis? What do the numbers on the vertical axis measure? What do the vertical bars on the graph show?

B. Use the first clue. Which bar shows a multiple of 3, 6, and 8?

C. Use the second clue. What are you looking for? Which pair of bars fits this clue? Which of those bars includes Dave's vote?

D. Find the bar that includes Peggy's vote. Which grade is each student in? Which issue did each student choose?

Read the problem again. Look at the information given and the main question. Review your work. Is your answer reasonable?

Use Logical Reasoning

39

At the Greer Road 4th of July picnic, 22 people ate hot dogs, 52 ate burgers, and 37 had potato salad. 5 people had ONLY potato salad. 10 people had hot dogs and potato salad. Some people had burgers and potato salad. 15 people didn't eat hot dogs, potato salad, or burgers. How many people came to the picnic?

① FIND OUT

A. What is the question you have to answer?

B. How many people ate hot dogs? Burgers? Potato salad? Of the people who had potato salad, how many had ONLY potato salad? How many people had hot dogs as well as potato salad?

C. How many people didn't have hot dogs, potato salad, or burgers?

② CHOOSE A STRATEGY

I can _________________________________ to solve the problem.

       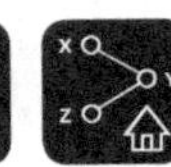  

A. What does each circle stand for? What belongs in space F? What label could you give space B? Space D? Do we know how many people belong in either space?

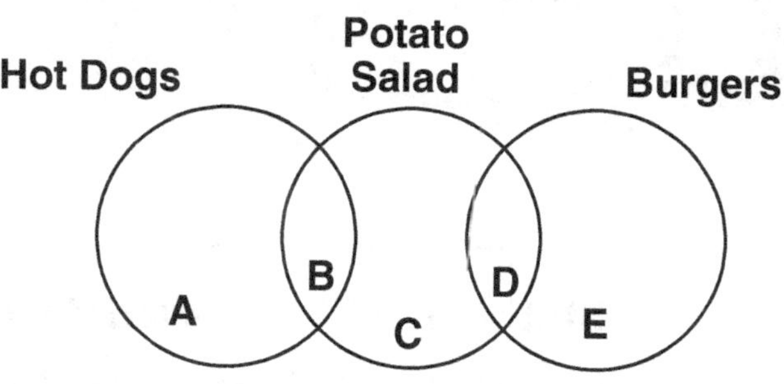

F =

B. How many people ate hot dogs? Does this include the number in B? Then how can you find the number for A?

C. How many people ate potato salad? How many ate only potato salad? Where does that number belong? Then how can you find the number of people who ate both potato salad and burgers?

D. Find the number of people who ate only burgers. Then how can you find the total number of people who came?

Read the problem again. Look at the information given and the main question. Review your work. Is your answer reasonable?

Use Logical Reasoning

Station KQST surveyed 310 people about
how they got their news. There were
130 people who used the Internet, 86 who
read newspapers, and 160 who watched TV.
45 people used the Internet and newspapers,
38 people used the Internet and TV, and
50 people used newspapers and TV. 25 people
used all three ways. How many people did not
use any of the three ways to get news?

40

1 FIND OUT

A. What is the question you have to answer?

B. How many people did KQST survey? How
many people used the Internet? Newspapers?
TV? Internet and newspapers? Internet and TV?
Internet, TV, and newspapers?

2 CHOOSE A STRATEGY

I can _________________________ to solve the problem.

       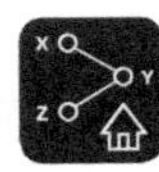  

A. How could you label space B? D? E? F? What number belongs in E?

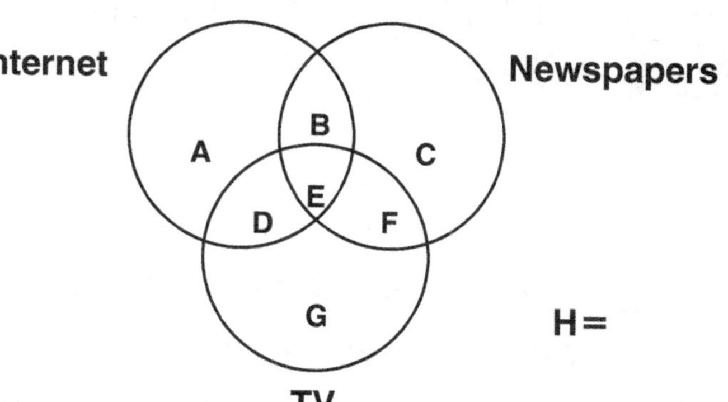

B. 45 people used the Internet and newspapers. Does that include the people who used all three ways? How can you find the number who used just the Internet and newspapers? Put the number in space B.

C. 38 people used the Internet and TV. Does that include the people who used all three ways? Find the number for D.

D. If you have numbers for B, D, and E, then how can you find the number of people who used only the Internet?

E. Now find the number for space F. Then fill in C and G.

F. How can you find the total number of people who use any of the three ways? How can you find the number of people who don't use any of the three ways?

Read the problem again. Look at the information given and the main question. Review your work. Is your answer reasonable?

Make It Simpler

41

The Chinese New Year's parade is ready to begin. Fifteen judges are stationed along the parade route at 10-yard intervals. Where should the judges meet after the parade, so that all together they walk the shortest possible distance?

① FIND OUT

A. What is the question you have to answer?

B. How many judges are stationed along the parade route?

C. How far apart are the judges stationed?

D. What is the condition for the location of the judges' meeting place?

② CHOOSE A STRATEGY

I can _________________________ to solve the problem.

A. What does the first diagram show?

B. What are the possible meeting places for 2 judges? Which is the best meeting place? Why?

C. What are the possible meeting places for 3 judges? Which is the best meeting place? How far would each judge walk? How many yards would they walk in all?

D. What are the possible meeting places for 4 judges? Which is the best? How far would each judge walk? How many yards would they walk in all?

E. Now try making a diagram for 15 judges. What is the best place for them to meet, so that all together they travel the fewest possible yards?

4 LOOK BACK

Read the problem again. Look at the information given and the main question. Review your work. Is your answer reasonable?

Make It Simpler

42

In the museum's collection of **85,800** antique toys, there are one-half as many dolls as toy cars, one-third as many cars as teddy bears, and one-fourth as many bears as wind-up tin animals. If each doll is worth at least **$43.50**, each tin animal is worth at least **$12.50**, each bear is valued at no less than **$22.50**, and each car is worth **$30.00** or more, what is the minimum value of the collection?

1 FIND OUT

A. What is the question you have to answer?

B. How many dolls are there? How many cars? How many teddy bears? How many toys are there in all?

C. What is the minimum value of a doll? A car? A teddy bear? A tin animal?

2 CHOOSE A STRATEGY

I can _________________________________ to solve the problem.

       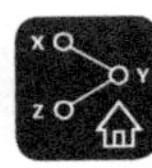  

A. Let's work with a total of 858 instead of 85,800.

B. Look at the way the numbers of toys are related to each other. If you make a guess for the right kind of toy, you'll be able to find a number for every other kind of toy. Which kind of toy will you make a guess for?

C. What is your guess? Then how many of each other kind of toy are there? How can you check your guess? Compare your guess to the simpler total we're working with. Was your guess too high or too low? Keep guessing until you have the right number of toys.

D. How will you find the minimum value of the collection? What is the minimum value of the collection of 858 toys?

E. Compare the actual total to our simpler total. How many times greater is the actual number of toys? How can you adjust your answer to fit a total of 85,800 toys?

Read the problem again. Look at the information given and the main question. Review your work. Is your answer reasonable?

Work Backwards

43

The Space Board was taking applications for places in the city being built in space. Half of the applications came from Kenya, $\frac{1}{4}$ from Japan, $\frac{1}{8}$ from India, $\frac{1}{16}$ from Poland, $\frac{1}{32}$ from the United States, and the rest came from Brazil. If Brazil sent 18 applications, how many applications were received, and how many came from each country?

A. What are the questions you have to answer?

B. How many of the applications came from Kenya? From Japan? From India? From Poland? From the United States? From Brazil?

I can _________________________________ to solve the problem.

     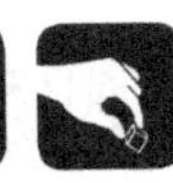  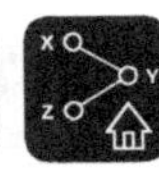  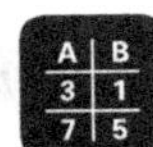

A. How can you label the left half of the diagram? How should you divide the diagram to show Japan's fraction of the applications? India's? Poland's? The United States'? What fraction of the applications came from Brazil? What number can we write in Brazil's section?

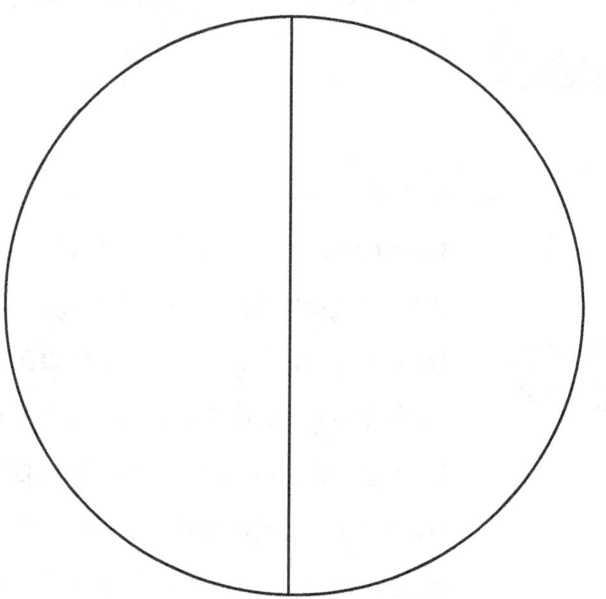

B. Now work backwards on the diagram. What section is equal to Brazil's section? Then how many applications came from that country? Write the number. Which section is equal to Brazil's and the United States' sections together? Then how many applications came from that country? Keep working backwards on the diagram by finding equal parts. In each section, write the number of applications from that country.

C. How many applications were received, and how many came from each country?

Read the problem again. Look at the information given and the main question. Review your work. Is your answer reasonable?

Work Backwards

44

Ramón, a bicycle messenger, picked up some packages this morning. His instructions were to drop off $\frac{1}{2}$ of those packages at the Bryant building and pick up 2; then to drop off $\frac{1}{2}$ of the packages at the hospital and pick up 7; to go to City Hall next, drop off $\frac{1}{4}$ of the packages and pick up 1; to deliver $\frac{1}{2}$ of the packages and pick up 1 at a law office on his fifth stop; and to deliver the 6 packages he had left on the sixth stop. How many packages did Ramón deliver?

1 FIND OUT

A. What is the question you have to answer?

B. What did Ramón do on his first stop? His second? His third? His fourth? His fifth? His sixth?

2 CHOOSE A STRATEGY

I can _________________________ to solve the problem.

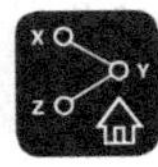

Stop 1	Stop 2	Stop 3	Stop 4	Stop 5	Stop 6
					A = 6 D = 6 C = 0 P = 0

A = packages he arrived with
D = packages delivered
C = packages he is still carrying
P = packages picked up

A. Begin with Ramón's sixth stop. Did he pick up any packages? Did he keep any packages to carry to the next stop? How many did he deliver there? How many did he have when he arrived at the sixth stop?

B. Now you know he left his fifth stop with 6 packages. What was the last thing he did on this stop? How many packages was he carrying before he picked it up? What did he do before picking it up? If he had 5 packages after delivering half, then how many did he have when he arrived at his fifth stop?

C. Work backwards to his fourth stop.

D. Work backwards. How many packages did Ramón deliver?

Read the problem again. Look at the information given and the main question. Review your work. Is your answer reasonable?

A
B
C

Make It Simpler

45

Your sock drawer contains 24 yellow socks, 30 blue striped socks, 17 orange socks, 12 white socks, 34 pale purple socks, 30 royal red socks, 11 green spotted socks, 14 black socks, and 23 brown socks. If you reach into the drawer in the dark, how many socks do you need to pull out to be sure you have a matching pair?

1 FIND OUT

A. What is the question you have to answer?

B. How many yellow socks are in the drawer? How many blue? How many orange? How many white? How many purple? How many red? How many green? How many black? How many brown?

2 CHOOSE A STRATEGY

I can _________________________ to solve the problem.

       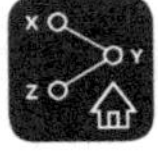  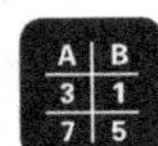

A. Begin with just 2 colors—yellow and blue—and just 2 socks of each color. If you pull out a yellow sock first, which socks are still available for your second selection? Is it certain that the second sock you pull out will match the sock in your hand?

B. If the second sock you pull is blue, you have 1 yellow and 1 blue. Which socks are available for your third selection? Is it certain that the third sock you pull out will match one of the socks in your hand?

C. Now try 3 colors—yellow, blue, and orange—and just 2 socks of each color. Do you see a pattern? Will the pattern hold even if there are more than 2 socks of each color?

D. Now consider the original problem. Use the pattern you found in the simpler problems. How many socks do you need to pull out to be sure you have a matching pair?

4 LOOK BACK

Read the problem again. Look at the information given and the main question. Review your work. Is your answer reasonable?

Make It Simpler

46

The mechanic reads in the log of the Pegasus that it has been 4,740,000 minutes since the hovercraft's last maintenance check at 4 P.M. on April 12, 2026. What is the time and date of the present maintenance check?

① FIND OUT

A. What is the question you have to answer?

B. How many minutes have passed since the Pegasus' last maintenance check?

C. When was the last maintenance check?

② CHOOSE A STRATEGY

I can ___________________________ to solve the problem.

         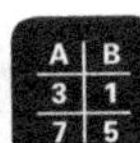

A. Let's start with only 474 minutes. How can you divide up the minutes? How would you find the date of the present maintenance check?

B. Now try 4,740 minutes. How can you divide up this number? Can you divide the number further? How would you find the date of the present maintenance check?

C. Now use the number of minutes in the original problem. What will you do first? Next? Is there a way to divide this number further?

D. Does every year have the same number of days? Calculate the number of years without accounting for leap years. If the year 2000 was a leap year, how many leap years would occur in that period? Now adjust your answer to account for leap years.

E. Count on from April 12, 2026 to find out the present maintenance check time and date.

4 **LOOK BACK**

Read the problem again. Look at the information given and the main question. Review your work. Is your answer reasonable?

Brainstorm

47

Ana presented a problem to Luis: "If you can solve this," said Ana, "I'll buy you the ice cream cone of your choice! Here's the problem: Show how one-third of six is one." Luis got his ice cream cone. What was Luis' answer?

1 FIND OUT

A. What is the question you have to answer?

B. What problem did Luis solve?

2 CHOOSE A STRATEGY

I can _________________________ to solve the problem.

        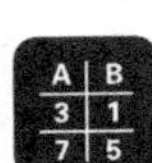

A. What is your first reaction to this problem?

B. If you set up a visual picture of this problem, what do you see?

C. What different kinds of visual pictures can you think of to represent six?

D. Write down all the things you can think of. Try playing with all the pictures you come up with by dividing each in thirds in different ways. Is there some way to do this and come up with one?

4 LOOK BACK

Read the problem again. Look at the information given and the main question. Review your work. Is your answer reasonable?

Brainstorm

48

Long after everyone else was in bed, Mark still sat at the table. He was leaning intently over the geometric shapes he was creating with toothpicks. Mark was trying to make 6 squares with 12 toothpicks. Can you help him?

1 FIND OUT

A. What is the question you have to answer?

B. What is Mark trying to make?

C. How many toothpicks does Mark have?

2 CHOOSE A STRATEGY

I can _________________________ to solve the problem.

A. What is your first reaction to Mark's project?

B. If Mark needs 4 toothpicks to make 1 square, then how many do you think he would need to make 6 squares?

C. Can he make 6 *separate* squares with only 12 toothpicks? Then how can he reduce the number of toothpicks from 24?

D. Do all the squares have to be flat on the table?

E. How can you make 6 squares with 12 toothpicks?

4 LOOK BACK

Read the problem again. Look at the information given and the main question. Review your work. Is your answer reasonable?

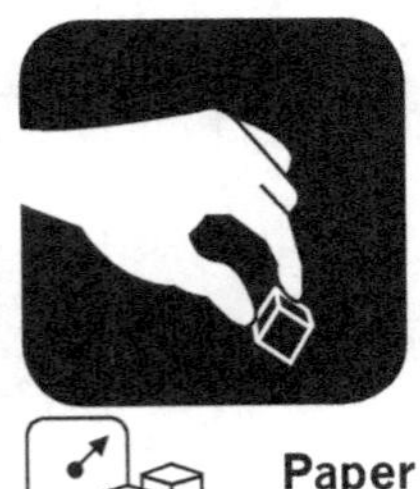

Act Out or Use Objects

49

Lily, her sister Lulu, Lily's boyfriend Lee, and his brother Louis are riding a float in the parade. One person faces outward on each side of the float. Lulu is to Lee's girlfriend's left. Lee's brother is to Lily's boyfriend's right. Each Anderson is across from a Jones. There is not an Anderson on the west side. Louis is on the north. What is each person's full name and position?

1 FIND OUT

A. What is the question you have to answer?

B. What are the positions? Where is Louis? Lulu? Lee's brother?

C. What do you know about each family?

2 CHOOSE A STRATEGY

I can ______________________________ to solve the problem.

A. Since each person has 2 names, a first and last, that you have to match, how many scraps of paper do we need? How will you label each scrap?

B. Draw a square to represent the float. How can you label the positions?

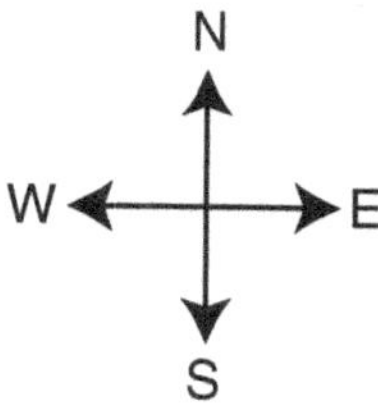

C. Begin with the person whose position you know. Which name can you place on the diagram? What else do you know about that person's position? Then where must that person be?

D. What do you know about where Lulu is sitting? Place Lulu on one side of the float. Who has to be on her right? If there is no room at Lulu's right, move Lulu to another position.

E. What clue do you have about the Andersons and the Joneses?

F. Move your papers around until you find a solution.

G. What is each person's full name and position?

Read the problem again. Look at the information given and the main question. Review your work. Is your answer reasonable?

Act Out or Use Objects

50

The Kings, the Montoyas, the Hansons, and the Cohens—four married couples—are playing a card game. Partners sit across from each other. Married couples are not at the same table. Joshua is sitting opposite Alex's wife. Paul's wife is on Joshua's left and across from Mary's husband. Joshua's partner is a Montoya. Martin's wife is on Suzanne's right. Mary's partner is not a King. No King is across from a Cohen or a Montoya. Mary is across from Jen's husband. Joshua's wife is to Mary's left and across from Anna's husband. What is each person's full name, and where is each person sitting?

1 FIND OUT

A. What is the question you have to answer?

B. Find out what the problem tells you.

2 CHOOSE A STRATEGY

I can _______________________ to solve the problem.

       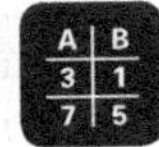

A. How many scraps do you need to represent the first and last names of the 8 card players? How will you label each scrap?

B. Draw squares to represent the tables.

C. Begin with Joshua. Place him at one of the tables. What are all the things you know about Joshua?

D. If Joshua is at the first table, then where is Joshua's wife? What else do you know about Joshua's wife?

E. Continue moving names around on the diagram to fit the clues. What is each person's full name, and where is each person sitting?

4 LOOK BACK

Read the problem again. Look at the information given and the main question. Review your work. Is your answer reasonable?

Use Logical Reasoning

51

Five cousins are camping together. Two are Pattersons, and three are Andersons. Two are from Montana, and three are from Wyoming. Irene and Yvonne are from the same state. Michael and Emily are from different states. Leo and Michael have the same last name. Emily and Yvonne do not have the same last name. The Patterson from Montana awakes in the middle of the night to discover a raccoon rummaging through the food pack. What is that cousin's first name?

1 FIND OUT

A. What is the question you have to answer?

B. What are the cousins' names? How many are Pattersons? How many are Andersons? How many are from Montana? How many are from Wyoming?

2 CHOOSE A STRATEGY

I can _________________________________ to solve the problem.

       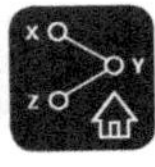  

Patterson	Anderson	Montana	Wyoming
Emily/Yvonne	Emily/Yvonne		

A. How many names will you have to write under Montana? Under Wyoming? How many names will you have to write under Patterson? Under Anderson?

B. If Leo and Michael have the same last name, and Emily and Yvonne have different last names, then how many people have Leo and Michael's last name? Under which last name would you write each of their first names? Under which last name would you write Irene's name?

C. If Irene and Yvonne are from the same state, and Michael and Emily are from different states, then under which state would you write each of their first names? Under which state would you write Leo's name?

D. What is the first name of the Patterson from Montana who discovered the raccoon?

Read the problem again. Look at the information given and the main question. Review your work. Is your answer reasonable?

Use Logical Reasoning

52

Seven members of the school orchestra have been selected to participate in the All-City Orchestra. They are Caleb, Paul, Patty, Ricky, Jesse, Miranda, and Ann. Three of them are 13 years old, and four of them are 12 years old. Three play the trumpet, and four play the violin. Caleb, Paul, and Ricky play the same instrument. Jesse and Miranda play different instruments. Jesse, Miranda, and Ann are the same age. Caleb and Paul are not the same age. What is the name of the 13-year-old who plays trumpet?

❶ FIND OUT

A. What is the question you have to answer?

B. What are musicians' names? What are their ages? Their instruments?

❷ CHOOSE A STRATEGY

I can _______________________ to solve the problem.

       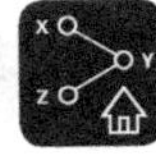  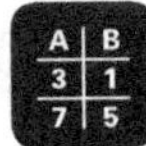

Age 13	Age 12	Plays Trumpet	Plays Violin
Paul/Caleb	Paul/Caleb		

A. How many names will you have to write under Age 13? Under Age 12?

B. How many names will you have to write under Trumpet? Under Violin?

C. If Caleb, Paul, and Ricky play the same instrument, and Miranda and Jesse play different instruments, then how many people play Caleb's instrument? Under which instrument would you write each of their names? Then under which instrument would you write Ann's and Patty's names?

D. Fill in the names below Age 13 and Age 12.

E. What is the name of the 13-year-old trumpet player?

4 LOOK BACK

Read the problem again. Look at the information given and the main question. Review your work. Is your answer reasonable?

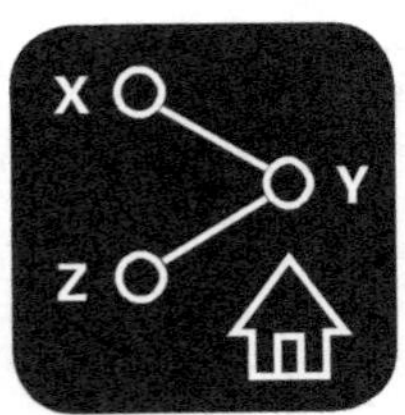

Use or Make a Picture or Diagram

53

Nikko is arranging photographs for a page in the school yearbook. The five pictures that he is working with are all the same size square. Each photo has to share at least one side with another photo. What are all the different ways Nikko could place the pictures on the page?

❶ FIND OUT

A. What is the question you have to answer?

B. How many photos is Nikko trying to arrange on the page?

C. What shape are the photographs? What do you know about the sizes of the photos?

D. How must the photos be arranged?

❷ CHOOSE A STRATEGY

I can _________________________ to solve the problem.

A. What is the least number of sides each photo must share with another photo?

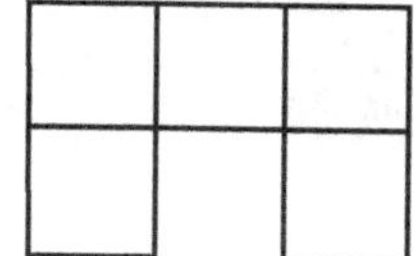

B. Draw the squares or shade in squares on graph paper to show one way that Nikko could arrange the five photos.

C. Is there another way Nikko could arrange the pictures?

D. What are all the different ways Nikko could place the pictures on the page?

Read the problem again. Look at the information given and the main question. Review your work. Is your answer reasonable?

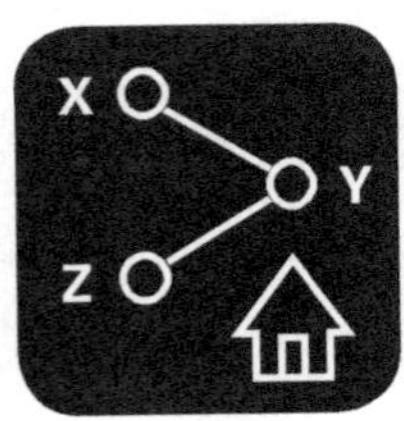

Use or Make a Picture or Diagram

54

King Hoptuma called his advisors together and said, "I want to build a structure that will tell all the world of my greatness. I want it to be made up of six triangular pieces of gold. Each triangle must be equilateral, and share at least one side with another triangle. All triangles must be the same size. I hereby command you to present to me as many different designs as possible." With that, King Hoptuma's advisors scurried off. What are all the different designs the advisors could present to the king?

1 FIND OUT

A. What is the question you have to answer?

B. How many pieces of gold must the advisors use? What shape must the pieces be? What size? How must the gold triangles be arranged?

2 CHOOSE A STRATEGY

I can ___________________________ to solve the problem.

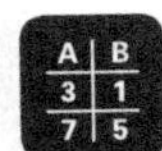

A. How many sides must each triangle share with another triangle? Draw the triangles or shade in triangles on graph paper to show one design that the advisors would present to the king.

B. Is there another way the advisors could arrange the triangles?

C. Can you think of a way to organize your arrangements to be sure you make as many as possible?

4 LOOK BACK

Read the problem again. Look at the information given and the main question. Review your work. Is your answer reasonable?

Guess and Check

55

"Seven-eighths of the number, added to 72, doubles the number," read Soren. "Is this a joke?" Larry looked at the paper and said, "No, it's the number of my locker." What is the number of Larry's locker?

1 FIND OUT

A. What is the question you have to answer?

B. What do you know about the number of the locker?

2 CHOOSE A STRATEGY

I can _________________________ to solve the problem.

       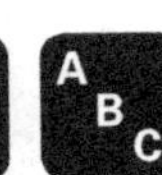 

A. Since you have to find $\frac{7}{8}$ of a number, begin with numbers that are easy to divide in eighths. What number's multiples will you begin with?

B. What is your guess? What is $\frac{7}{8}$ of that number? What number do you get if you add 72? Is that double your guess?

C. If your guess was wrong, how can you use that information to make your next guess better? Continue guessing and checking until you find a solution.

D. What is the number of Larry's locker?

4 LOOK BACK

Read the problem again. Look at the information given and the main question. Review your work. Is your answer reasonable?

Guess and Check

56

The Split Ends are hot! They have played 20 live shows this month, and every record store in town is sold out of their CDs. Their manager told reporters that over a thousand CDs were sold in one hour after the last concert. He said, "If you take $\frac{4}{6}$ of the number and double it, that's 500 more than the number of CDs sold." Give the reporters a break. How many CDs were sold?

1 FIND OUT

A. What is the question you have to answer?

B. What do you know about the number of CDs sold?

2 CHOOSE A STRATEGY

I can _______________________ to solve the problem.

         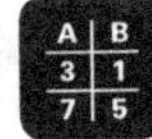

A. Since you have to take $\frac{4}{6}$ of your guess, what number's multiples should you use for your first guesses? What is your guess?

B. What is $\frac{4}{6}$ of the number you guessed? What is the result when you double that? What is the result when you subtract 500 from that? Is that the same number as your guess?

C. If your guess was wrong, how can you use the information to make your next guess? Continue guessing and checking until you find a solution.

D. How many CDs were sold?

4 LOOK BACK

Read the problem again. Look at the information given and the main question. Review your work. Is your answer reasonable?

57

Kent and José are playing a game with polyhedral dice, one die numbered 1 to 6 and the other numbered 1 to 8. They roll both dice and then multiply one number by the other. If the product is divisible by 5, Kent gets 5 points. If the product is divisible by 6, José gets 5 points. The first player to get 30 points is the winner. Is this game fair or unfair? Why?

① FIND OUT

A. What are the questions you have to answer?

B. What are Kent and José doing?

C. What numbers are on one die? What numbers are on the other die?

D. How do the players score points?

② CHOOSE A STRATEGY

I can ________________________ to solve the problem.

       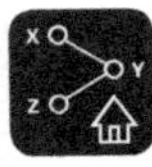  

A. What numbers are on one die? Does each number have an equal chance of coming up?

B. What numbers are on the other die? Does each number have an equal chance of coming up?

x	1	2	3	4	5	6	7	8
1	1	2	3	4	5	6	7	8
2								
3								
4								
5								
6								

C. What will you keep track of in the first row of the table? What will you keep track of in the rest of the rows? Fill in the rest of the table.

D. How many of the products are divisible by 5? How many of the products are divisible by 6?

E. Is this game fair or unfair? Why?

Read the problem again. Look at the information given and the main question. Review your work. Is your answer reasonable?

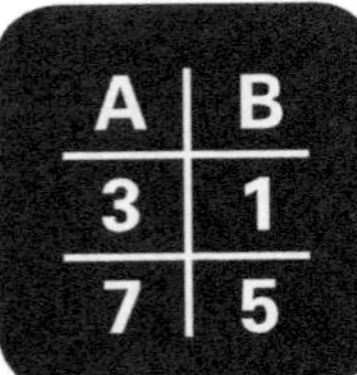

Use or Make a Table

58

Marta and Amy are playing a game. Each player rolls a number cube, numbered 1–6. They use the numbers rolled to make a fraction, with Marta's number as the numerator and Amy's number as the denominator. Marta gets 2 points if the fraction is a whole number greater than 1. Amy gets 2 points if the fraction can be simplified to a mixed number. Is the game fair or unfair? Why?

1 FIND OUT

A. What are the questions you have to answer?

B. What kind of number cubes are Marta and Amy using for their game? How do they use the numbers rolled?

C. How do they earn points in the game?

2 CHOOSE A STRATEGY

I can _________________________ to solve the problem.

       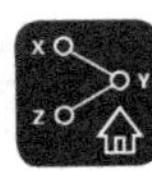  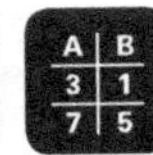

A. What numbers are on each number cube? Does each number have an equal chance of being rolled?

B. Look at the table that has been started. Fill in all the possible fractions that can be rolled.

Fractions with Numbers Rolled

Number Rolled On Marta's Cube		Number Rolled on Amy's Cube					
		1	2	3	4	5	6
	1	$\frac{1}{1}$					
	2	$\frac{2}{1}$					
	3	$\frac{3}{1}$					
	4						
	5						
	6						

C. How many fractions are whole numbers greater than 1? How many fractions can be simplified to a mixed number?

D. Is the game fair or unfair? Why?

Read the problem again. Look at the information given and the main question. Review your work. Is your answer reasonable?

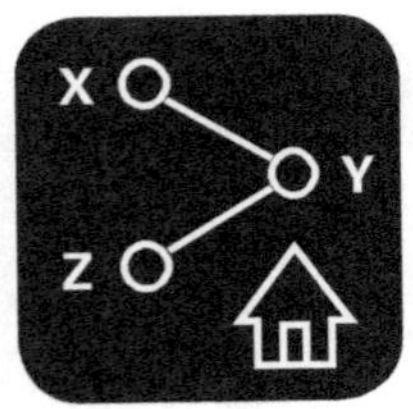

Use or Make a Picture or Diagram

59

Sally's class did a survey of bags of peanuts. They counted the contents of 35 bags and recorded their data in a line plot.

- **Sally's bag had 2 less than the mode.**
- **Jan's bag had 4 more than the mean.**
- **Warren's bag had a square number.**

How many peanuts did each of the students count?

1 FIND OUT

A. What is the question you have to answer?

B. How did the students conduct their survey? How many bags of peanuts did they have?

C. What do you know about the number in each student's bag?

2 CHOOSE A STRATEGY

I can _________________________ to solve the problem.

③ SOLVE IT

```
                            X
                            X
                            X
                   X        X
                   X        X                       X
         X    X    X        X    X                  X
         X    X    X        X    X                  X
         X    X    X        X    X         X        X
    X    X    X    X        X    X         X        X
   ─────────────────────────────────────────────────────
    48   49   50   51       52   53   54   55   56   57   58
```

Number of Peanuts in a Bag

A. What is shown along the bottom of the graph? What do the Xs above each value show?

B. What is the first clue? What do you have to find out first? How can you find it? What number is it? What was Sally's total?

C. What is the next clue? How can you find the mean? What number is it? What was Jan's total?

D. What will you look for as you use the last clue? What was Warren's total?

E. How many peanuts did each of the students count?

④ LOOK BACK

Read the problem again. Look at the information given and the main question. Review your work. Is your answer reasonable?

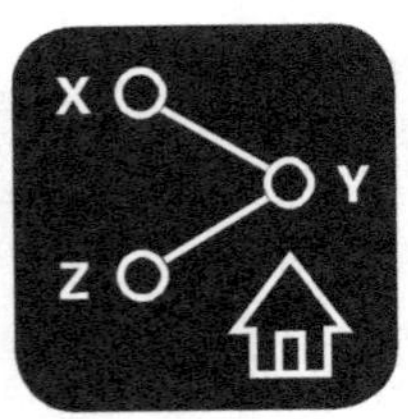

Use or Make a Picture or Diagram

60

The sixth and seventh graders made a stem-and-leaf plot of their travel times to and from school. In the sixth grade, Julio takes 30 minutes more than the mode, and Andreas takes 11 minutes more than the median. In the seventh grade, James takes 10 minutes more than the mean, and Angelina takes 20 minutes more than the median. How long does it take each of these students to travel to and from school?

1 FIND OUT

A. What is the question you have to answer?

B. What does the graph show?

C. What do you know about Julio and Andreas? About James and Angelina?

2 CHOOSE A STRATEGY

I can _________________________________ to solve the problem.

    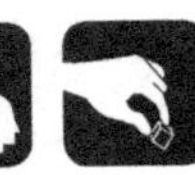    

3 SOLVE IT

Travel Time To and From School

6th grade ones	tens	7th grade ones
8 5 2 0	1	2 5 6 8
8 6 5 0 0 0 0	2	2 2 5 5 8 8
6 5 4 0 0	3	0 0 0 4 5 8
8 5 5 0 0 0	4	2 5 5 6 8
5 4 0 0	5	3 4 5 5
0	6	0 0 5 6

Key

0 | 1 | 2 means Grade 6 = 10 minutes,
Grade 7 = 12 minutes.

A. To find out how long it takes Julio, what do you have to look for first? What is it? How long does it take Julio?

B. How can you find the median? What is the median for sixth grade? How long does it take Andreas?

C. To find out how long it takes James, what do you have to find first? How can you find it? How long does it take James?

D. Continue in the same way for Angelina.

4 LOOK BACK

Read the problem again. Look at the information given and the main question. Review your work. Is your answer reasonable?

Use or Look for a Pattern

61

Benito used centimeter cubes to make a square array in which each cube touched the sides of the neighboring cubes. He kept adding cubes until each edge of the square array was **8 centimeters** long. How many faces of cubes could be seen on Benito's largest array, and how many faces of cubes were hidden?

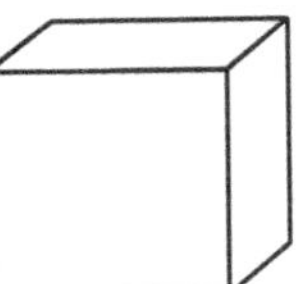

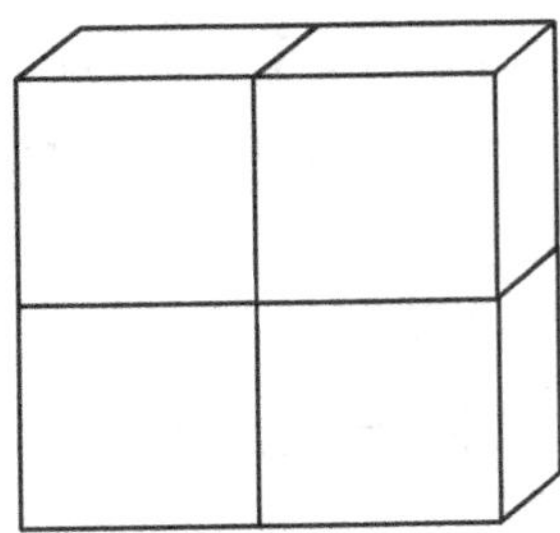

① FIND OUT

A. What is the question you have to answer?

B. What did Benito build with the centimeter cubes? How large was the final array?

C. How long is each edge of 1 centimeter cube? How many faces does a centimeter cube have?

② CHOOSE A STRATEGY

I can _________________________________ to solve the problem.

A. How many faces does a centimeter cube have? If you placed the cube on a table, how many faces could you see? How many faces are hidden?

B. In all, how many cubes would you need to build the next larger array? How many faces of cubes could you see? How many cube faces would be hidden?

C. Fill in the table with the numbers for the next array. How many cubes are there? How many faces show? How many are hidden?

Cubes in Array	Cube Faces Seen	Cube Faces Hidden
1	5	1
4		
9		

D. Do you see a pattern in the way the number of faces seen changes? Do you see a pattern in the numbers of hidden faces? Use the patterns to complete your table. What is the largest array that you need to include in your table?

E. How many faces of cubes could be seen on Benito's largest array, and how many were hidden?

Read the problem again. Look at the information given and the main question. Review your work. Is your answer reasonable?

Use or Look for a Pattern

62

Jenna drew a 1-by-1 rhombus, a 2-by-2 rhombus, and a 3-by-3 rhombus. She connected dots to divide each drawing into the maximum number of rhombuses. The rhombuses could overlap, but no triangles could appear. Jenna found a pattern. How many rhombuses will she find in a 12-by-12 rhombus?

① FIND OUT

A. What is the question you have to answer?

B. What is a 1-by-1 rhombus? What other rhombuses did Jenna draw?

C. What did Jenna look for to find a pattern? What conditions did she set for how she drew the lines?

② CHOOSE A STRATEGY

I can _________________________ to solve the problem.

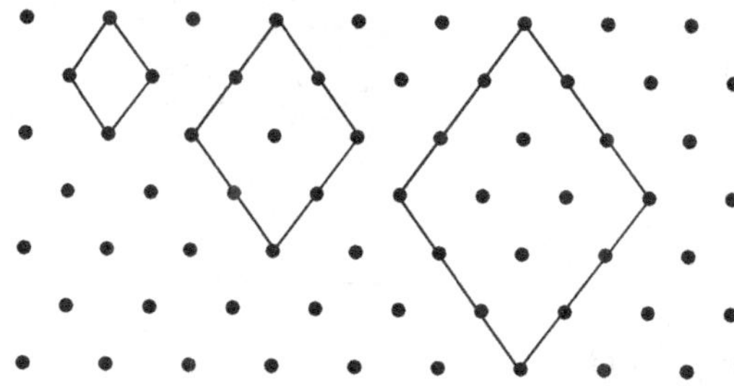

Drawing	Rhombuses in Each Drawing			
	1-by-1	2-by-2	3-by-3	4-by-4
1-by-1	1			
2-by-2	4	1		
3-by-3	9	4	1	
4-by-4				

A. How many rhombuses are in a 1-by-1 rhombus?

B. How many 1-by-1 rhombuses are in a 2-by-2 rhombus? How many 2-by-2 rhombuses are there?

C. How many 1-by-1 rhombuses are in a 3-by-3 rhombus? How many 2-by-2 rhombuses are there? 3-by-3?

D. Can you see a pattern in the numbers of your table? What do you notice about those numbers?

E. How many rhombuses will she find in a 12-by-12 rhombus?

4 LOOK BACK

Read the problem again. Look at the information given and the main question. Review your work. Is your answer reasonable?

Make It Simpler

63

Thirteen friends are planning their visits to one another's homes. Each friend is going to spend a week with each other friend. If each friend sees each other friend at least once, what is the smallest number of days in all that the friends will spend visiting each other?

1 FIND OUT

A. What is the question you have to answer?

B. How many friends are in the group? How do the friends visit each other? If friend A visits friend B in B's home, then does B also have to go to A's home?

C. How long does a friend spend at another friend's home?

2 CHOOSE A STRATEGY

I can _________________________ to solve the problem.

      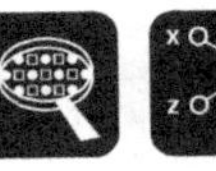 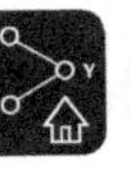  

 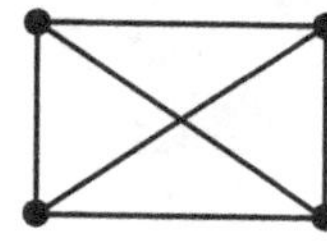

Number of Friends	Number of Visits
2	1
3	3
4	
5	

A. Begin with 2 friends. How many visits will be made?

B. How many visits will be made by 3 friends? By 4 friends? By 5 friends?

C. Do you see a pattern in the way the numbers of visits increase? What is the pattern? Use the pattern and continue filling in your table until you have numbers for all 13 friends.

D. How long is each visit? What is the smallest number of days in all that the friends will spend visiting each other?

4 LOOK BACK

Read the problem again. Look at the information given and the main question. Review your work. Is your answer reasonable?

Make It Simpler

64

Plant A produces $\frac{1}{2}$ as many batches of corn-based plastic containers per hour as Plant B, which produces $\frac{1}{2}$ as many as Plant C, which produces $\frac{1}{2}$ as many as Plant D. Last year all four plants together produced 262,800,000 batches of containers. The profit on each batch is $25. If the plants operate around the clock, how much profit was made per hour at each plant?

1 FIND OUT

A. What is the question you have to answer?

B. How many batches of containers were produced at Plant A? B? C? How many batches were made last year in all four plants?

C. How much profit did the company make per batch?

2 CHOOSE A STRATEGY

I can _______________________ to solve the problem.

 SOLVE IT

A. Try 262,800 in place of 262,800,000. If 262,800 batches were produced per year and if the plants operate around the clock, how many batches did the plants produce per hour?

B. Which plant produced the most batches of containers? The fewest?

C. If you make a guess about the number of batches produced by one plant in an hour, how can you figure out how many batches each other plant produces per hour? How many did each plant produce each hour?

D. Now try 262,800,000. How many batches of containers were produced per hour by each plant?

E. How much profit was made on each batch? How much profit was made per hour at each power plant?

 LOOK BACK

Read the problem again. Look at the information given and the main question. Review your work. Is your answer reasonable?

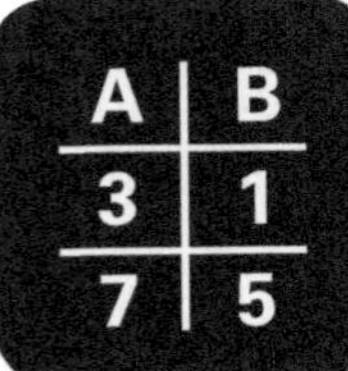

Use or Make a Table

65

Professors Orion and Polaris work at the observatory every night. On Monday, during his 8 P.M. to 10 P.M. shift, Professor Orion notices 12 new stars. He finds 18 on Tuesday, 30 on Wednesday, 36 on Thursday, and 48 on Friday. Professor Polaris, working from 10 P.M. to midnight, notices 18 new stars on Monday, 26 on Tuesday, 44 on Wednesday, 52 on Thursday, and 70 on Friday. If the number of stars each professor sees keeps increasing in the same way, on which day will the total number of new stars pass 365?

1 FIND OUT

A. What is the question you have to answer?

B. How many stars does Professor Orion notice each day? Professor Polaris?

2 CHOOSE A STRATEGY

I can _________________________ to solve the problem.

       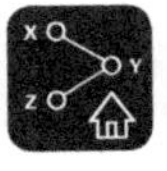  

Days	M	T	W	T	F			
Orion	12	18	30					
Polaris	18	26						
Total	30							

A. What will you keep track of in the Orion row? In the Polaris row? In the last row? Fill in the numbers for the first 5 days.

B. How did Professor Orion's numbers change from Monday to Tuesday? From Tuesday to Wednesday? From Wednesday to Thursday? From Thursday to Friday? Do you see a pattern?

C. How did Professor Polaris's numbers change from Monday to Tuesday? From Tuesday to Wednesday? From Wednesday to Thursday? From Thursday to Friday? Do you see a pattern?

D. Continue to fill in your table. Find the totals as you work. On which day of the week will the total number of new stars pass 365?

Read the problem again. Look at the information given and the main question. Review your work. Is your answer reasonable?

Use or Make a Table

66

Biologists have made an amazing discovery at a certain watering hole. The animals have varied their schedules in order to prevent confrontations with other animals. Look at the table to see the animals that the biologists counted in the first four days. They count 11 wildebeests, 8 elephants, 4 zebras, and 9 lions on the fifth day. If these patterns continue, how many animals will the biologists have counted all together by the end of the first day that they don't see any zebras?

1 FIND OUT

A. What is the question you have to answer?

B. Find out what the problem tells you.

2 CHOOSE A STRATEGY

I can _________________________ to solve the problem.

       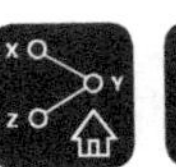  

Day	1	2	3	4	
Wildebeest	7	10	9	12	
Elephant	4	2	6	4	
Zebra	6	8	5	7	
Lion	5	4	7	6	
Total					

A. What will you keep track of in the rows?

B. How did the number of wildebeests change from the first day to the second day? The second to the third? The third to the fourth? The fourth to the fifth? What is the pattern?

C. How did the number of elephants change from the first day to the second day? The second to the third? The third to the fourth? The fourth to the fifth? What is the pattern?

D. Find the pattern for each kind of animal. Continue to fill in your table. How will you know when your table is complete?

E. How many animals will the biologists have counted all together by the end of the first day that they don't see any zebras?

4 LOOK BACK

Read the problem again. Look at the information given and the main question. Review your work. Is your answer reasonable?

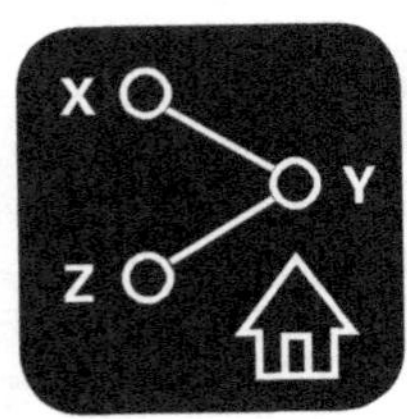

Use or Make a Picture or Diagram

67

Alexis and Rachel are playing a game. They take turns tossing 3 pennies in the air. If 3 heads or 3 tails come up, then Rachel gets 9 points. If 2 heads or 2 tails come up, then Alexis gets 3 points. The first player to get 24 points is the winner. Is this game fair or unfair? Why?

① FIND OUT

A. What are the questions you have to answer?

B. How do the players score points?

C. Who wins the game?

② CHOOSE A STRATEGY

I can _________________________ to solve the problem.

        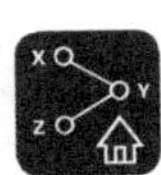  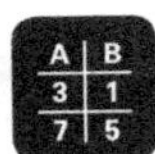

A. What are the possible outcomes of the first flip? Start with the tree that shows H for the first flip. What are the 2 possibilities for the second flip? If the second penny showed heads, what possibilities should we show below that for the third flip? Trace the left-hand branches of the diagram to find one outcome of the three flips. What is that outcome? What other outcome can you find by tracing the branches we've completed so far?

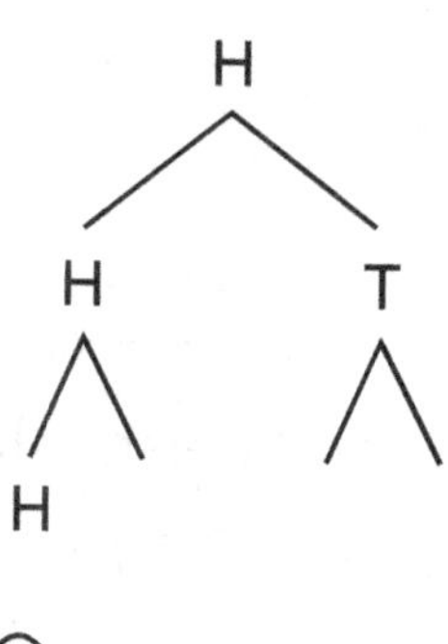

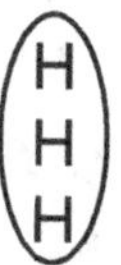

B. To complete this tree, go back to the top. If the first flip of heads was followed by a second flip of tails, what do we need to show next? What outcomes do these branches lead to?

C. Fill in the branches of the other tree diagram.

D. In all, how many outcomes are possible? How many show 3 heads or 3 tails? How many show 2 heads or 2 tails? Think about how the girls earn points and how many points they earn for each of the possible outcomes. Is this game fair or unfair? Why?

Read the problem again. Look at the information given and the main question. Review your work. Is your answer reasonable?

Use or Make a Picture or Diagram

68

Janna and Riley are playing a game. They put 3 green cubes, 3 blue cubes, and 3 red cubes in a bag and take turns drawing 2 cubes at a time out of the bag. After each turn they return the cubes to the bag. If the cubes they draw are the same color, then Janna gets 5 points. If the cubes are not the same color, then Riley gets 5 points. The first player to get 25 points is the winner. Is this game fair or unfair? Why?

❶ FIND OUT

A. What are the questions you have to answer?

B. What cubes are Janna and Riley using? What do the players do on each turn? After each turn?

C. How do the players score points? Who wins?

❷ CHOOSE A STRATEGY

I can _______________________________ to solve the problem.

A. Think of the three green cubes as G1, G2, and G3. What will you call the other cubes?

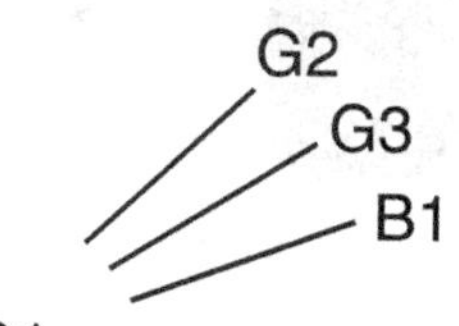

B. The first diagram will show all the possible pairs that could have G1 in them. Which cubes are in the first pair? The second pair? The third pair? Now fill in the rest of the pairs that have G1 in them.

C. That diagram included a pair with G2. The next diagram will show the other possible pairs that have G2 in them. What are they?

D. What will the next diagram show? Make the rest of the diagrams.

E. How many outcomes are possible? How many pairs have 2 cubes of the same color? How many pairs have 2 cubes of different colors? Is this game fair or unfair? Why?

4 LOOK BACK

Read the problem again. Look at the information given and the main question. Review your work. Is your answer reasonable?

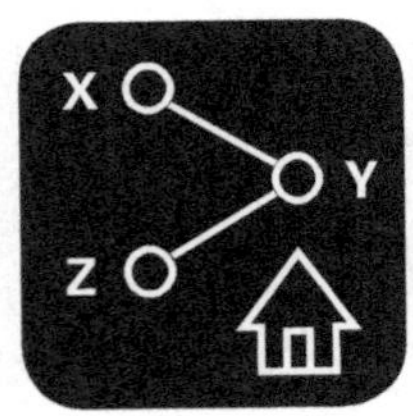

Use or Make a Picture or Diagram

69

Ellen and Lauren are covering a rectangular prism with special paper. The prism is 20 inches long, 8 inches wide, and 6 inches high. They must cut out a one-piece net that fits the prism exactly, with no gaps or overlapping. What are the dimensions of the smallest sheet of paper from which they can cut the net?

1 FIND OUT

A. What is the question you have to answer?

B. What are Ellen and Lauren making?

C. What are the dimensions of the rectangular prism?

D. What are the conditions for cutting out the net?

2 CHOOSE A STRATEGY

I can _______________________________ to solve the problem.

A. Finish the diagram of the rectangular prism and add measurement labels. How long is the prism? How wide? How high?

B. Now draw a diagram of the net the girls are making. Which side of the prism will you cover first?

C. The next side you cover must be connected to the first. Which side of the prism will you cover next?

D. How will you cover the small faces on the sides of the prism? Continue adding to the net until it is complete.

E. How will you find the dimensions of the smallest sheet of paper from which the girls can cut the net?

4 LOOK BACK

Read the problem again. Look at the information given and the main question. Review your work. Is your answer reasonable?

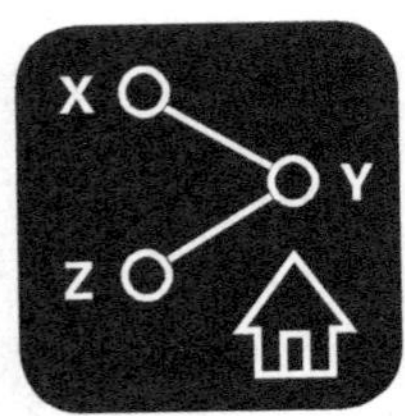

Use or Make a Picture or Diagram

70

Malcolm and his dad are buying colorful blocks of concrete for a garden patio. The space they plan to cover is shaped like this, with the length of the north side being equal to the shortest distance between the north and south sides. The blocks they can buy have different shapes, but each block will cover an area equal to one square foot. How many blocks will Malcolm and his dad need?

① FIND OUT

A. What is the question you have to answer?

B. What shape is the space they plan to cover? What are the measurements of the space? Of the angles inside the space?

C. How much of the space will each concrete block cover?

② CHOOSE A STRATEGY

I can _________________________________ to solve the problem.

       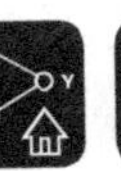  

A. What kind of angles are the two that are labeled?

B. How can you break the shape of the patio up into simpler shapes? Draw lines to separate those shapes. Use the fact about the north side of the patio. What is the shape at the center of the diagram?

C. Which shape's area should you find first? How can you find it?

D. How can you find the dimensions of the central shape in order to find its area?

E. How many square yards must Malcolm and his dad cover? How many square feet are in 1 square yard? How many blocks will Malcolm and his dad have to buy?

4 LOOK BACK

Read the problem again. Look at the information given and the main question. Review your work. Is your answer reasonable?

Thinking Questions
Questions to think about as you are solving problems

FIND OUT

What is the problem about?

What question do I have to answer?

What do I have to find out to solve the problem?

Are there any words or ideas I don't understand?

What information can I use?

Am I missing any information that I need?

CHOOSE A STRATEGY

Have I solved a problem like this before?

What strategy helped me solve it?

Can I use the same strategy for this problem?

SOLVE IT

What information should I start with?

Do I need to add, subtract, multiply, or divide?

How can I organize the information that I use or find?

Is the strategy I chose helpful?

Would another strategy be better?

Do I need to use more than one strategy?

Is my work easy to read and understand? Is it complete?

LOOK BACK

Did I answer the question that was asked in the problem?

Is more than one answer possible?

Is my math correct?

Does my answer make sense?

Can I explain why I think my answer is correct?